PROGRESS IN NICKEL TOXICOLOGY

PROGRESS IN NICKEL TOXICOLOGY

Proceedings of the
Third International Conference on
Nickel Metabolism and Toxicology
held in Paris 4–7 September 1984

edited by

STANLEY S. BROWN
Dudley Road Hospital, Birmingham, UK

and

F. WILLIAM SUNDERMAN Jr
University of Connecticut Medical School, USA

PUBLISHED FOR THE
INTERNATIONAL UNION OF
PURE AND APPLIED CHEMISTRY BY
BLACKWELL SCIENTIFIC PUBLICATIONS
OXFORD LONDON EDINBURGH
BOSTON PALO ALTO MELBOURNE

Published by Blackwell Scientific Publications
Editorial offices:
Osney Mead, Oxford, OX2 0EL
8 John Street, London, WC1N 2ES
23 Ainslie Place, Edinburgh, EH3 6AJ
52 Beacon Street, Boston
Massachusetts 02108, USA
667 Lytton Avenue, Palo Alto
California 94301, USA
107 Barry Street, Carlton
Victoria 3053, Australia

First published 1985

Printed in Great Britain at the University Press, Cambridge

DISTRIBUTORS

USA and Canada
Blackwell Scientific Publications Inc
PO Box 50009, Palo Alto
California 94303

Australia
Blackwell Scientific Book Distributors
31 Advantage Road, Highett
Victoria 3190

British Library
Cataloguing in Publication Data
International Conference on Nickel Metabolism and Toxicology *(3rd: 1984: Paris)*
Progress in nickel toxicology: proceedings of the third International Conference on Nickel Metabolism and Toxicology, held in Paris, 4–7 September 1984.
1. Nickel—Toxicology
I. Title II. Brown, Stanley, S.
III. Sunderman, F. William (Fredrick William), *1931–* IV. International Union of Pure and Applied Chemistry
615.9′25625 RA1231.N44

ISBN 0-632-01355-9

CONTENTS

CLINICAL PATHOLOGY

CONTRIBUTORS AND SESSION CHAIRMEN

ADAM, SAFIA, *Laboratoire de Spectroscopie Biomoleculaire, U.E.R. de Médécine, 74 rue Marcel Cachin, 93012 Bobigny, France*

AITIO, ANTERO, *Institute of Occupational Health, SF-00290 Helsinki, Finland*

AL-RABAEE, RAZZAK N., *Cytology Unit, Departments of Botany and Zoology, University of Manchester, Manchester, M13 9PL, UK*

BENSON, JANET M., *Lovelace Inhalation Toxicology Research Institute, Albuquerque, New Mexico 87185, USA*

BERNABEO, FERRUCCIO, *Occupational Health Unit, Hospital of Lecco, 22053 Lecco, Italy*

BERTHOLF, ROGER L., *Depts of Pathology, Biochemistry & Internal Medicine, University of Virginia Medical Center, Charlottesville, Virginia 22908, USA*

BISHOP, CAROLE MARY, *Health and Safety Executive, 25 Chapel Street, London NW1 5EE, UK*

BLANC, MICHEL, *INSERM U 88-GERSS, 91 bd de l'Hôpital, F-75634 Paris Cedex 13, France*

BOURBONNAIS, RENEE, *INSERM U 88-GERSS, 91 bd de l'Hôpital, F-75634 Paris Cedex 13, France*

BOURTAYRE, PIERRE, *Laboratoire de Spectroscopie Biomoleculaire, U.E.R. de Médécine, 74 rue Marcel Cachin, 93012 Bobigny, France*

BOYSEN, MORTEN, *Departments of Pathology and Biochemistry, The Norwegian Radium Hospital and Norsk Hydro's Institute for Cancer Research, and Ear, Nose and Throat Department, Rikshospitalet, Oslo, Norway, and the Norwegian Cancer Society*

BOZEC, CHRISTIAN, *Societé le Nickel, Tour Maine Montparnasse, 33, avenue de Maine, 75755 Paris Cédex 05, France*

BRODEUR, JEAN-MARC, *INSERM U 88-GERSS, 91 bd de l'Hôpital, F-75634 Paris Cedex 13, France*

BROWN, STANLEY S., *Regional Laboratory for Toxicology, Dudley Road Hospital, Birmingham B18 7QH, UK*

BROWN, SUE, *Depts of Pathology, Biochemistry & Internal Medicine, University of Virginia Medical Center, Charlottesville, Virginia 22908, USA*

BRUNI, CARLO, *Department of Neurology, University of Virginia Medical Center, Charlottesville, Virginia 22908, USA*

CATALANATTO, FRANK A., *Department of Pediatric Dentistry, University of Connecticut Health Center, Farmington, CT 06032, USA*

CHANNON, SUSAN M., *Department of Medicine, Medical School, Framlington Place, Newcastle upon Tyne NE2 4HH, UK*

CHASTANG, JEAN-FRANCOIS, *INSERM U 88-GERSS, 91 bd de l'Hôpital, F-75634 Paris Cedex 13, France*

CHRISTENSEN, J. MOLIN, *Danish National Institute of Occupational Health, DK-2900 Hellerup, Denmark*

CIRLA, ANGELO M., *Occupational Health Unit, Hospital of Lecco, 22053 Lecco, Italy*

COLLYN D'HOOGHE, MAUD, *INSERM U-124, 1 Place de Verdun, 59045 Lille-Cedex, France*

COSTA, MAX, *University of Texas, Health Science Centre, Medical School, Division of Toxicology, P.O. Box 20708, Houston, Texas, USA*

COURBON, P., *Centre d'Etudes et Recherches des Charbonnages de France, B.P. No. 2, 60550 Verneuil-en-Halatte, France*

CRISOSTOMO, CRISTINA, *Department of Laboratory Medicine, University of Connecticut Health Center, Farmington, CT 06032, USA*

CROSS, JACQUELINE E., *Department of Applied Chemistry, UWIST, P.O. Box 13, Cardiff CF1 3XF, UK*

DAB, WILLIAM, *INSERM U 88-GERSS, 91 bd de l'Hôpital, F-75634 Paris Cedex 13, France*

DALEN, HELGE, *Laboratory of Clinical Electron Microscopy, University of Bergen, Bergen, Norway*

DAUTZENBERG, B., *Société Métallurgique Le Nickel-SLN, 75755 Paris Cedex 15, France*

DECOULX, J., *Service de Traumatologie, 1 Place de Verdun, 59045 Lille-Cedex, France*

DEWALLY, DANYS, *Laboratoire de Chimie Minerale Appliquée, Université des Sciences et Techniques de Lille, 59655 Villeneuve d'Ascq, France*

DRAZNIOWSKY, MICHAEL, *Department of Medicine, Medical School, Framlington Place, Newcastle upon Tyne NE2 4HH, UK*

DUNNICK, JUNE K., *National Toxicology Program, National Institute of Environmental Health Sciences, Research Triangle Park, NC 27709, USA*

EL-MASRY, HISHAM A., *Cytology Unit, Departments of Botany and Zoology, University of Manchester, Manchester M13 9PL, UK*

FRANCOIS, DENIS, *INSERM U 88-GERSS, 91 bd de l'Hôpital, F-75634 Paris Cedex 13, France*

FRONING, MARTINA, *Institute of Applied Physical Chemistry, Nuclear Research Center (KFA), Juelich, P.O. Box 1913, D-5170 Juelich, FRG*

FUHRER, REBECCA, *INSERM U 88-GERSS, 91 bd de l'Hôpital, F-75634 Paris Cedex 13, France*

GALEN, MALCOLM, *Department of Medicine, St Francis Hospital, Hartford, CT 06015, USA*

GILMAN, J.P.W., *Department of Biomedical Sciences, University of Guelph, Guelph, Ontario, Canada*

GODARD, CATHERINE, *INSERM U 88-GERSS, 91 bd de l'Hôpital, F-75634 Paris Cedex 13, France*

GOLDBERG, MARCEL, *INSERM U 88-GERSS, 91 bd de l'Hôpital, F-75634 Paris Cedex 13, France*

GOLDBERG, PAQUERETTE, *INSERM U 88-GERSS, 91 bd de l'Hôpital, F-75634 Paris Cedex 13, France*

GOLDWASSER, EUGENE, *Department of Biochemistry and Molecular Biology, University of Chicago Medical School, Chicago, IL 60637, USA*

GUSTAFSSON, TOM, *Institute of Occupational Health, SF-00290 Helsinki, Finland*

GYLSETH, BJORN, *Institute of Occupational Health, P.O. Box 9149 Dep, Oslo 1, Norway*

HAAPA, KAUKO, *A. Ahlstrom Oy, SF-48600 Karhula, Finland*

HANSEN, KAREN, *Danish National Institute of Occupational Health, Baunegaardsvej 73, DK-2900 Hellerup, Denmark*

HANSTEEN, INGER-LISE, *Department of Occupational Medicine, Telemark Sentralsjukehus, Porsgrunn, Norway*

HARATAKE, JOJI, *School of Medicine, University of Occupational and Environmental Health, 1-1 Iseigaoka, Yahatanishiku, Kitakyushu 807, Japan*

HAUGEN, AAGE, *Department of Toxicology, National Institute of Public Health, 0462 Oslo 4, Norway*

HENDERSON, ROGENE F., *Lovelace Inhalation Toxicology Research Institute, Albuquerque, New Mexico, 87185, USA*

HERLANT-PEERS, MARIE-CLAIRE, *INSERM U. 124, 1 Place de Verdun, 59045 Lille-Cedex, France*

HILDEBRAND, HARTMUT F., *Institut de Médecine du Travail, 1 Place de Verdun, 59045 Lille-Cedex, France*

HOCHRAINER, DIETER, *Fraunhofer-Institut für Toxikologie und Aerosolforschung, 5948 Schmallenberg-Grafschaft, FRG*

HOPFER, SIDNEY M., *Department of Laboratory Medicine, University of Connecticut Medical School, Farmington, CT 06032, USA*

HORIE, AKIO, *School of Medicine, University of Occupational and Environmental Health, Yahatanishi, Kitakyushu 807, Japan*

HUANG, CHARLES Y., *Laboratory of Biochemistry, National Heart, Lung and Blood Institute, National Institutes of Health, Bethesda, MD 20205, USA*

HUGHES, DINAH M., *Department of Applied Chemistry, UWIST, P.O. Box 13, Cardiff CF1 3XF, UK*

ISHIMATSU, SHIGEKO, *School of Medicine, University of Occupational and Environmental Health, Yahatanishi, Kitakyushu 807, Japan*

JACOBSEN, JORUNN, *Departments of Pathology and Biochemistry, The Norwegian Radium Hospital and Norsk Hydro's Institute for Cancer Research, and Ear, Nose and Throat Department, Rikshospitalet, Oslo, Norway, and the Norwegian Cancer Society*

JAMESON, C.W., *National Toxicology Program, National Institute of Environmental Health Sciences, Research Triangle Park, NC 27709, USA*

JARVISALO, JORMA, *A. Ahlstrom Oy, SF-48600 Karhula, Finland*

JASMIN, GAETAN, *Department of Pathology, Faculté de Médecine, Université de Montreal, Montreal, Canada H3C 3J7*

JULIAN, JAMES A., *Health Sciences Centre, McMaster University, Hamilton, Ontario, L8N 3Z5 Canada*

KAISER, ROLF-DIETER, *University of Dortmund, Department of Chemistry, Otto-Hahn-Strasse, D-4600 Dortmund 50, FRG*

KASPRZAK, KAZIMIERZ S., *Nutrition and Metabolism Section, Laboratory of Comparative Carcinogenesis, Division of Cancer Etiology, National Cancer Institute, Frederick, MD 21701, USA*

KERR, DAVID N.S., *Royal Postgraduate Medical School, Hammersmith Hospital, London W12 0HS, UK*

KIILUNEN, MIRJA, *Institute of Occupational Health, SF-00290 Helsinki, Finland*

KING, MARITA M., *Laboratory of Biochemistry, National Heart, Lung and Blood Institute, National Institutes of Health, Bethesda, MD 20205, USA*

KIRCHHOFF, M., *Department of Internal Medicine, C. Glostrup Hospital, University of Copenhagen, DK-2600 Glostrup, Denmark*

KLEIN, F., *Laboratoire d'Etude et de Controle de l'Environnement Siderurgique, 57210-Maizieres-les-Metz, France*

KLOCKOW, DIETER, *University of Dortmund, Department of Chemistry, Otto-Hahn-Strasse, D-4600 Dortmund 50, FRG*

KODAMA, YASUSHI, *School of Medicine, University of Occupational and Environmental Health, 1-1 Iseigaoka, Yahatanishiku, Kitakyushu 807, Japan*

KRESS, WOLFRAM, *Institute of Occupational and Social Medicine, University of Erlangen-Nurnberg, Schillerstrasse 25, D-8520 Erlangen, FRG*

LEACH, CHARLES N. Jr. *Department of Medicine, New Britain General Hospital, New Britain, CT 06050, USA*

LECLERC, ANNETTE, *INSERM U88-GERSS, 91 bd de l'Hôpital, F-75634 Paris Cedex 13, France*

LINDEN, JEANNE V., *Department of Laboratory Medicine, University of Connecticut Health Center, Farmington, CT 06032, USA*

LIQUIER, JEAN, *Laboratoire de Spectroscopie Biomoleculaire, U.E.R. de Médecine, 74 rue Marcel Cachin, 93012 Bobigny, France*

LUMB, GEORGE, *Department of Pathology, Hahnemann University, Philadelphia, Pennsylvania 19102, USA*

LYNN, KIRK K., *Laboratory of Biochemistry, National Heart, Lung and Blood Institute, National Institutes of Health, Bethesda, MD 20205, USA*

MARZOUK, ABUBAKR, *Departments of Laboratory Medicine and Pharmacology, University of Connecticut Medical School, Farmington, Connecticut 06032, USA*

MASTROMATTEO, ERNEST, *INCO Ltd, 1 First Canadian Place, Toronto, Ontario, M5X 1C4 Canada*

MATSUNO, KOJI, *School of Medicine, University of Occupational and Environmental Health, Yahatanishi, Kitakyushu 807, Japan*

McCLELLAN, ROGER O., *Lovelace Inhalation Toxicology Research Institute, Albuquerque, New Mexico 87185, USA*

McLEAN, JACK R., *Radiation Protection Bureau, Health and Welfare Canada, Ottawa, K1A 0L2 Canada*

MEININGER, JEAN, *Société Métallurgique Le Nickel-SLN, 75755 Paris Cedex 15, France*

MERCIER, J.M., *Institute de Médecine du Travail, 1 Place de Verdun, 59045 Lille-Cedex, France*

MIKAC-DEVIC, MIRNA, *University Hospital Dr. M. Stojanovic, Vinogradska 29, 41000 Zagreb, Yugoslavia*

MORGAN, LINDSAY G., *Medical Department, INCO Europe, Clydach, Swansea SA6 5QR, UK*

MUIR, DAVID C.F., *Health Sciences Centre, McMaster University, Hamilton, Ontario, L8N 3Z5 Canada*

NIEBOER, EVERT, *Department of Biochemistry, McMaster University, Hamilton, Ontario, L8N 3Z5 Canada*

NORSETH, TOR, *Institute of Occupational Health, P.O. Box 9149 Dep, Oslo 1, Norway*

NOVELLI, ETHELL BARBOSA, *Department of Biochemistry, Instituto Basi co de Biologia, Medica e Agricola, UNESP, Sao Paulo, Brazil*

NURNBERG, HANS WOLFGANG, *Institute of Applied Physical Chemistry, Nuclear Research Center (KFA), Juelich, P.O. Box 1913, D-5170 Juelich, FRG*

OLDIGES, HUBERT, *Fraunhofer-Institut für Toxikologie und Aerosolforschung, 5948 Schmallenberg-Grafschaft, FRG*

OSTAPCZUK, PETER, *Institute of Applied Physical Chemistry, Nuclear Research Center (KFA), Juelich, P.O. Box 1913, D-5170 Juelich, FRG*

OTTOBONI, FABRIZIO, *Lofarma Research Laboratory, Viale Cassala 40, 20143 Milano, Italy*

OZONAS, BARTOLOME RIBAS, *Department of Biochemistry, Consejo Superior de Investigaciones Cientificas, Universidad Complutense de Madrid, Facultat de Farmicia, 28003, Madrid, Spain*

PALMER REISS, ANE-LENE, *Institute of Occupational Health, P.O. Box 9149 Dep, Oslo 1, Norway*

PARKINSON, IAN S., *Department of Medicine, Medical School, Framlington Place, Newcastle upon Tyne NE2 4HH, UK*

POIRIER, LIONEL A., *Nutrition and Metabolism Section, Laboratory of Comparative Carcinogenesis, Division of Cancer Etiology, National Cancer Institute, Frederick, MD 21701, USA*

POON, TIMOTHY F-H., *Renal Unit, Princess Margaret Hospital, Kowloon, Hong Kong*

RAITHEL, HANS-JURGEN, *Institute of Occupational and Social Medicine, University of Erlangen-Nurnberg, Schillerstrasse 25, D-8520 Erlangen, FRG*

RATTI, RENATA, *Occupational Health Unit, Hospital of Lecco, 22053 Lecco, Italy*

REBAR, A.H., *Department of Veterinary Microbiology, Pathology and Public Health, School of Veterinary Medicine, Purdue University, W. Lafayette, Indiana 47907, USA*

REICHBORN-KJENNERUD, SISSEL, *The Norwegian Radium Hospital and Norsk Hydro's Institute for Cancer Research, Montebello, Oslo 3, Norway and The Norwegian Cancer Society*

REID, MARILYN C., *Departments of Laboratory Medicine and Pharmacology, University of Connecticut Medical School, Farmington, Connecticut 06032, USA*

REITH, ALBRECHT, *The Norwegian Radium Hospital and Norsk Hydro's Institute for Cancer Research, Montebello, Oslo 3, Norway and The Norwegian Cancer Society*

REUS, M. ISABEL SANCHEZ, *Department of Biochemistry, Consejo Superior de Investigaciones Cientificas, Universidad Complutense de Madrid, Facultad de Farmacia, 28003 Madrid, Spain*

RIGAUT, JEAN-PAUL, *URBB 263 INSERM, Université Paris 7, 2, Place Jussieu, Tour 53, 75251 Paris Cedex 05, France*

ROBERTS, ROBIN S., *Health Sciences Centre, McMaster University, Hamilton, Ontario, L8N 3Z5 Canada*

RODRIGUES, NEY LOBATO, *Department of Biochemistry, Instituto Basi co de Biologia, Medica e Agricola, UNESP, Sao Paulo, Brazil*

ROSS, RICHARD, *Depts of Pathology, Biochemistry & Internal Medicine, University of Virginia Medical Center, Charlottesville, Virginia 22908, USA*

ROUMAZEILLE, B., *Service de Traumatologie, 1 Place de Verdun, 59045 Lille-Cedex, France*

ROY, BERNARD R., *Corporate Director of Loss Prevention, AMAX Inc., AMAX Center, Greenwich, Connecticut 06836, USA*

SATRUSTEGUI, ICIAR BASAGOITI, *Department of Biochemistry, Consejo Superior de Investigaciones Cientificas, Universidad Complutense de Madrid, Facultad de Farmacia, 28003 Madrid, Spain*

SAVORY, JOHN, *Depts of Pathology, Biochemistry & Internal Medicine, University of Virginia Medical Center, Charlottesville, Virginia 22908, USA*

SCHALLER, KARL-HEINZ, *Institute of Occupational and Social Medicine, University Erlangen-Nurnberg, Schillerstrasse 25, D-8520, Erlangen, FRG*

SCHNEIDER, HENRY P., *Institute for Clinical Science, Hahnemann University, Philadelphia, Pennsylvania 19102, USA*

SEGNAN, NEREO, *INSERM U 88-GERSS, 91 bd de l'Hôpital, F-75634 Paris Cedex 13, France*

SIGEE, DAVID C., *Cytology Unit, Departments of Botany and Zoology, University of Manchester, Manchester M13 9PL, UK*

SJOHOLM, JAN, *Hoganas AB, Box 501, 5-26301 Hoganas, Sweden*

SKAUG, VIDAR, *Institute of Occupational Health, P.O. Box 8149 Dep, Oslo 1, Norway*

SMIALOWICZ, RALPH J., *Developmental Biology Division, Health Effects Research Laboratory, Research Triangle Park, North Carolina 27711, USA*

STERN, RICHARD M., *The Danish Welding Institute, DK-2600 Glostrup, Denmark*

STOEPPLER, MARKUS, *Institute of Applied Physical Chemistry, Nuclear Research Center (KFA), Juelich, P.O. Box 1913, D-5170 Juelich, FRG*

SUNDERMAN, F. WILLIAM Jr, *Depts of Laboratory Medicine and Pharmacology, University of Connecticut Medical School, Farmington, Connecticut 06032, USA*

SUNDERMAN, F. WILLIAM Sr, *Department of Pathology and Institute for Clinical Science, Hahnemann University, Philadelphia, Pennsylvania 19102, USA*

SWIERENGA, SABINE H.H., *Drug Toxicology Division, Health and Welfare Canada, Ottawa, K1A 0L2 Canada*

TABOURY JEAN A., *Laboratoire de Spectroscopie Biomoleculaire, U.E.R. de Médecine, 74 rue Marcel Cachin, 93012 Bobigny, France*

TAILLANDIER, ELIANE, *Laboratoire de Spectroscopie Biomoleculaire, U.E.R. de Médecine, 74 rue Marcel Cachin, 93012 Bobigny, France*

TAKENAKA, SHINJI, *Fraunhofer-Institut für Toxikologie und Aerosolforschung, Stadtfelddamm 35, 3000 Hannover, FRG*

TANAKA, ISAMU, *School of Medicine, University of Occupational and Environmental Health, 1-1 Iseigaoka, Yahatanishiku, Kitakyushu 807, Japan*

TAUTU, CONSTANTIN, *Department of Pathology, Faculté de Médecine, Université de Montreal, Montreal, H3C 3J7 Canada*

THOMASSEN, YNGVAR, *Institute of Occupational Health, P.O. Box 8149, 0033 Oslo, Norway*

THOMSEN, EBBE, *Oldvejen 7, Rogerup, DK-4050 Skibby, Denmark*
TONKS, DAVID, *Montreal General Hospital, 1650 Cedar Avenue, Montreal, Quebec, H3G 1A4 Canada*
TOSSAVAINEN, ANTTI, *Institute of Occupational Health, SF-00290 Helsinki, Finland*
TSUCHIYA, KENZABURO, *School of Medicine, University of Occupational and Environmental Health, 1-1 Iseigaoka, Yahatanishiku, Kitakyushu 807, Japan*
VALENTIN, HELMUT, *Institute of Occupational and Social Medicine, University Erlangen-Nurnberg, Schillerstrasse 25, D-8520, Erlangen, FRG*
VOSS, RAGNER, *Department of Pathology, The Norwegian Radium Hospital and Norsk Hydro's Institute for Cancer Research, Montebello, Oslo 3, Norway and The Norwegian Cancer Society*
WAGG, RICHARD M., *Health and Safety Executive, 403-405 Edgware Road, London NW2 6LN, UK*
WARD, MICHAEL K., *Department of Medicine, Medical School, Framlington Place, Newcastle upon Tyne NE2 4HH, UK*
WARNER, J. STUART, *INCO Ltd, 1 First Canadian Place, Toronto, Ontario, M5X 1C4 Canada*
WELTLE, DIETER, *Institute of Occupational and Social Medicine, University of Erlangen-Nurnberg, Schillerstrasse 25, D-8520 Erlangen, FRG*
WILLIAMS, DAVID R., *Department of Applied Chemistry, UWIST, P.O. Box 13, Cardiff CF1 3XF, UK*
WILLS, MICHAEL R., *Depts of Pathology, Biochemistry & Internal Medicine, University of Virginia Medical Center, Charlottesville, Virginia 22908, USA*
YAMASHIRO, SHIGETO, *Department of Biomedical Sciences, University of Guelph, Guelph, Ontario, Canada*
ZAHARIA, ODETTE, *Departments of Laboratory Medicine and Pharmacology, University of Connecticut Medical School, Farmington, Connecticut 06032, USA*
ZOBER, ANDREAS, *Institute of Occupational and Social Medicine, University Erlangen-Nurnberg, Schillerstrasse 25, D-8520, Erlangen, FRG*

PREFACE

The Third International Conference on Nickel Toxicology was held at the PLM St. Jacques Hotel in Paris, France, from 4 to 7 September 1984, under the joint sponsorship of the International Union of Pure and Applied Chemistry (IUPAC), the Association of Clinical Scientists, and the Nickel Producers Environmental Research Association (NiPERA). The Paris Conference was attended by 150 participants from 19 countries, including many of the world's authorities on nickel in the areas of trace analysis, biochemistry, radiochemistry, pharmacology, toxicology, pathology, immunology, industrial hygiene, epidemiology, occupational health, and clinical medicine. The text of the Richard T. Barton memorial lecture and synopses of the scientific papers that were presented at the Conference are published in this volume to facilitate rapid communication of progress in nickel toxicology.

The organization of the Paris Conference required the efforts of many scientists and physicians, including especially Christian Bozec (France), Danys DeWally (France), Ernest Mastromatteo (Canada), Lindsay Morgan (UK), Evert Nieboer (Canada), John Savory (USA), F. William Sunderman Sr (USA) and David Tonks (Canada). The authors gratefully acknowledge the valuable help of these and other colleagues; also generous support from the sponsoring organizations, and the assistance of Simon Rallison of Blackwell Scientific Publications Ltd in expediting the publication of these Proceedings.

Stanley S. Brown
Regional Toxicology Laboratory
Dudley Road Hospital
Birmingham UK

F. William Sunderman Jr
University of Connecticut
Medical School
Farmington CT USA

INTRODUCTION : 100 YEARS OF NICKEL TOXICOLOGY

F. WILLIAM SUNDERMAN Jr* and STANLEY S. BROWN**
President* and Vice-President** of the Clinical Chemistry Division, International Union of Pure and Applied Chemistry

NICKEL TOXICOLOGY 1884-1950

The Paris Conference celebrated the centennial of T. P. Anderson Stuart's pioneering research on nickel toxicity (Stuart, 1884). Although Gmelin (1826) had reported that salts of nickel and 16 other metals are toxic for animals, the science of nickel toxicology was established by the studies that Stuart (Fig. 1a) performed in 1880-83 at the Pharmacology Institute in Strasburg, Germany (now a part of Alsace-Lorraine in France). Oswald Schmeideberg (Fig. 1b), the Professor of Pharmacology at Strasburg, had been educated in medicine at Dorpat in Estonia (now Tartu in USSR) and he had served subsequently as assistant to Rudolph Buchheim, Professor of Pharmacology in Dorpat. In 1872, Schmeideberg accepted the newly created Chair of Pharmacology at Strasburg, and he soon attracted numerous young physicians and scientists for pre-doctoral and post-doctoral training at the Institute of Pharmacology (Fig. 2). Several of Schmeideberg's students became world-famous leaders of pharmacology, including John Jacob Abel, Otto Loewi, Arthur Cushny, and Torald Sollman (Holmstedt and Liljestrand, 1981; Fisher, 1984). Anderson Stuart, a medical graduate of the University of Edinburgh, joined Schmeideberg's staff in 1880 and began to study the toxicities of nickel and cobalt compounds. Previously, Schmeideberg had supervised other students in investigations of the toxicities of lead, platinum, antimony, mercury, and tin; accounts of their findings may be found in the *Archiv der Experimentellen Pathologie und Pharmacology*. In Stuart's own words (Epps, 1922) -

> "... I studied at Strasburg physiology and chemistry under Hoppe-Seyler, experimental physiology under Goltz, and experimental pharmacology under Schmeideberg. I used to also go round the medical wards with Kussmal. In Schmeideberg's laboratory I investigated the physiological properties of the salts of nickel and cobalt, and on the results wrote a thesis for my M.D.

FIG. 1 (a) T.P. Anderson Stuart - a photograph taken at the time of his pioneering work on nickel toxicology; (b) Oswald Schmeideberg Professor of Pharmacology at Strasburg.

FIG. 2 The Institute of Pharmacology, Strasburg, where T.P. Anderson Stuart's work on nickel toxicology was conducted.

> (obtained in 1882) for which I was awarded a gold medal .... Oscar Schmeideberg was a great authority on pharmacology, and was beloved by everyone; he was so kindly. gentle and anxious to tell his pupils all he knew, and it was a real pleasure to work under him. ...I had a very happy time at Strasburg; everyone was very kind to me, and I made friends who have remained friends ever since."

Stuart's report in Schmeideberg's *Archiv* (1884) provided the first comprehensive survey of the acute toxic effects of nickel salts, which were administered by oral and parenteral routes to frogs, pigeons, rats, guinea-pigs, rabbits, cats and dogs. The questions that Stuart posed concerning mechanisms of acute nickel toxicity are only now beginning to be answered by research on cellular uptake of nickel by calcium ion channels and by studies of free radical mechanisms for toxic injury of cells by nickel ions.

That Stuart was highly regarded by Schmeideberg is evident from correspondence appended to Epps' biography:-

> From Dr. O. Schmeideberg, Professor and Director of the Pharmacological Institute, Strasburg [Translation.]
>
> Strasburg, May 25th 1882.
>
> My Dear Anderson Stuart,
>
> You inform me of your intention of applying for the post of Teacher of Anatomy and Physiology at the Medical School lately founded in Sydney. On the one hand, I regret this, as it will take you so far away from us; on the other hand, I shall be glad if you find this an opportunity which will enable you to follow your inclination and devote yourself entirely to a scientific career. With your love and zeal for scientific labours and investigations as well as with your indefatigable endeavours to acquire the intellectual and manual accomplishments necessary for a Teacher and Investigator in the domain of the medical sciences, I became sufficiently acquainted when I had the pleasure of seeing you in my laboratory successfully engaged in the carrying on of scientific investigations. I know how well you succeeded in acquiring a profound knowledge and skill, and am convinced that, be the field of your activity in your native country or far away among the antipodes, we shall become indebted to you for many a valuable scientific discovery.
>
> I wish you may have the success you deserve, and am, your most devoted
>
> O. Schmeideberg.

In 1884, Stuart accepted the Chair of Physiology at the University of Sydney, Australia; he did not pursue further research on nickel toxicity, and his great work was the foundation, establishment and development of the Medical School at Sydney, where he died in 1920.

NICKEL TOXICOLOGY 1950 - 1984

During the 1950's and early 1960's, interest in nickel toxicology was rekindled by the researches of Richard Doll, John P.W. Gilman, William C. Hueper, Gaetan Jasmin, J. Gwynne Morgan, and F. William Sunderman, Sr (for references, see National Acadamy of Sciences,1975). The remarkable growth of research on nickel toxicology from 1960 to 1984 is attested by the annual citations on nickel in *Cumulative Index Medicus* (Fig. 3).

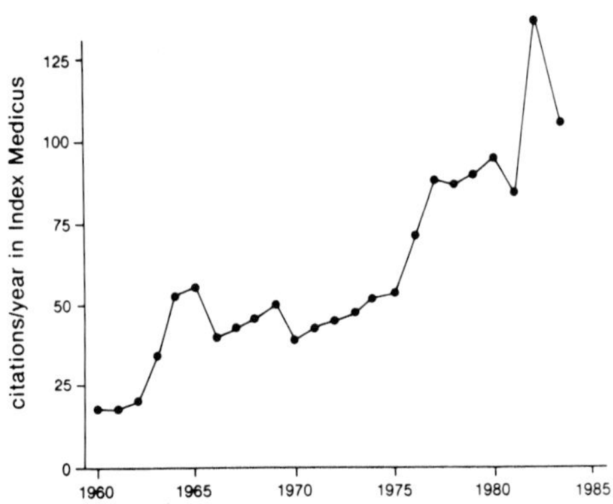

FIG. 3. Annual citations on nickel in *Cumulative Index Medicus* (1960-1984).

In 1975, the Commission on Toxicology, IUPAC Clinical Chemistry Division, established the Subcommittee on Environmental and Occupational Toxicology of Nickel, the founding members being -

I. Andersen (Nor)
R. T. Barton (Can)
R. Bourdon (Fr)
S. S. Brown (UK)
J. J. Clary (USA)
A. C. Hogetveit (Nor)
K. S. Kasprzak (Pol)
E. Mastromatteo (Can)
M. D. D. McNeely (Can)
J. Meininger (Fr)
M. Mikac-Devic (Yug)
L. Morgan (UK)
S. Nomoto (Jap)
T. Norseth (Nor)
L. Renzoni (Can)
B. Roy (USA)
J. Sjoholm (Swe)
M. Stoeppler (FRG)
F. W. Sunderman Jr (USA)
D. Tonks (Can)
H. Zachariasen (Nor)

Sadly, Dr. Richard T. Barton died in 1981; the Plenary Lecture of the Paris Conference on Nickel Toxicology was dedicated to him, in recognition of his scientific leadership, clinical expertise, and dedication to cooperation between academia and the nickel industry.

The Paris Conference was the twelfth meeting of the Nickel Subcommittee:-

1	INCO Ltd. London, June 1976	Founding Meeting
2	Connecticut Univ. Farmington, Dec. 1976	Scientific Program
3	Congress Centre, Monte Carlo, Mar. 1977	COMTOX-1 Symposium
4	Warsaw Univ. Warsaw, Aug. 1977	IUPAC Assembly
5	Caledonian Hotel, Kristiansand, May 1978	1st Ni Conference
6	Congress Centre, Davos, Sept. 1979	IUPAC Assembly
7	University College Swansea, Sept. 1980	2nd Ni Conference
8	Leuven Univ. Leuven, Sept. 1981	IUPAC Assembly
9	Trinity College, Dublin, June 1982	Scientific meeting
10	Sheraton Hotel, Montreal, July 1982	COMTOX-2 Symposium
11	Technical Univ. Lyngby, Aug. 1983	IUPAC Assembly
12	St. Jacques Hotel, Paris, Sept. 1984	3rd Ni Conference

For background information on nickel analysis, metabolism, toxicology and health effects, readers may consult recent monographs and articles by Nriagu (1980), Brown and Sunderman (1980), Bencko (1983), Mushak (1984), Norseth and Piscator (1984) and Sunderman (1984).

ACKNOWLEDGEMENTS

Fig. 1a and Fig. 1b are reproduced respectively, with permission, from Epps (1922) and Holmsted and Liljestrand (1981). Fig. 2 is reproduced, by courtesy of Dr. James W. Fisher, from his article in Pharmacologist, 1984, 26, 31-36.

REFERENCES

Bencko V. (1983) J. Hyg. Epidemiol. Microbiol. Immunol. 27, 237-247.

Brown S.S. and Sunderman F.W. Jr. (Eds.) (1980) Nickel Toxicology, Academic Press, London.

Epps, W. (1922) Anderson Stuart, M.D. - Physiologist, Teacher, Builder, Organizer, Citizen. Angus & Robertson, Sydney.

Fisher J.W. (1984) Pharmcologist, 26, 31-36.

Gmelin C.G. (1826) Bull. Sci. Med. 7, 110-117.

Holmsted B. and Liljestrand G. (1981) Readings in Pharmacology, Raven Press, New York.

Mushak P. (1984) Clin. Lab. Annual 3, 249-269.

National Acadamy of Sciences USA, Report (1975) Nickel (Report on Environmental Pollutants.) NAS Printing and Publishing Office, Washington DC.

Norseth T. and Piscator M. (1984) In Vouk V. et al. (Ed.) Handbook on the Toxicology of Metals, North Holland Biomedical Press, Amsterdam (in press).
Nriagu J.O. (Ed.) (1980) Nickel in the Environment. John Wiley, New York.
Stuart T.P.A. (1884) Arch. Exp. Pathol. Pharmakol. 18, 151-173.
Sunderman F.W. Jr. (Ed.-in-Chief) (1984) Nickel in the Human Environment. International Agency for Research on Cancer, Lyon (in press).

OPENING ADDRESS:

The Richard T. Barton Memorial Lecture

MUSCLE TUMORIGENESIS BY NICKEL COMPOUNDS

J. P. W. GILMAN and SHIGETO YAMASHIRO
Department of Biomedical Sciences,
University of Guelph, Guelph, Ontario, Canada

INTRODUCTION

It was over 30 years ago that Hueper (1952) reported a 50 mg intrafemoral injection of metallic nickel powder to have induced a 25% sarcoma incidence in rats. Included in these tumours were several rhabdomyosarcomas along the injection route. Subsequently, Hueper (1955) also reported the induction of fibrosarcomas from subcutaneous implantation of metallic nickel in rabbits but not in mice. Hueper emphasized that nickel and certain other metallic dusts constituted an industrial cancer hazard in the refining industry. As a result of these observations, the established view that industrial cancer risk from nickel was limited to nickel carbonyl $Ni_2(CO)_4$ began to be queried (Eckardt 1959).

Our own investigations into nickel carcinogenesis began in 1959 when we received a sample of dust from the calcining and sintering rooms of a Port Colborne, Ontario nickel refinery which had had an excessively high incidence of upper respiratory tract and lung cancers among that exposure group of its employees. The major components of this refinery dust are shown in Table I.

TABLE I. Composition of refinery dust

Cupric oxide (CuO)	3.4 per cent
Nickel sulphate ($NiSO_4.6\ H_2O$)	20.0 " "
Nickel sulphide (Ni_3S_2)	57.0 " "
Nickel oxide (NiO)	6.3 " "
Cobalt oxide (CoO)	1.0 " "
Ferric oxide (Fe_2O_3)	1.8 " "
Silicon dioxide (SiO_2)	1.2 " "
Miscellaneous	2.0 " "
Moisture	7.3 " "

Extensive and prolonged animal inhalation studies which were undertaken elsewhere on the same dust and its components, produced essentially negative results (Moon 1969). However, in our laboratory initial intramuscular implantation of the crude refinery dust (Gilman and Ruckerbauer 1962) resulted in a high incidence of sarcomas in rats and mice.

AFFINITY OF NICKEL COMPOUNDS TO MUSCLE

Tests on nickel subsulphide and nickel oxide showed both these consituents of the dust to be tumorigenic to rats and mice (Table II). While both compounds gave rise to a preponderance of rhabdomyosarcomas in the rat, this distribution was reversed in mice, with many of the mouse tumours being undifferentiated sarcomas that were difficult to definitely classify. In the rats, Ni_3S_2 showed a somewhat greater tumorigenic affinity to muscle (86%) than did NiO (70%).

TABLE II. Rhabdomyosarcoma incidences - Ni_3S_2 and NiO

Species	Compound (Dose, mg)	% Tum.	# for Histology	Distrib. % Rhab.	Others
Mouse	Ni_3S_2 (5)	60	21	10	90
Rat	Ni_3S_2 (10)	83	22	86	14
Mouse	NiO (5)	66	29	17	83
Rat	" (10)	69	20	70	30

Subsequent to these early studies on metallic nickel, nickel subsulphide and nickel oxide, numerous other nickel compounds have been tested for their carcinogenicity. Some of these have been listed and reviewed in the IARC monograph on nickel and nickel compounds (1976). This review generally supported the evidence presented by Payne (1964) and from our laboratory (Gilman 1965) for an inverse relationship between the solubility of nickel compounds and their tumorigenicity.

The incidence of sarcomas at the site of injection of six nickel compounds are listed in Table III and were the subject

TABLE III. Carcinogenicity of Ni Compounds[1]

Compound	Numb. Rats	% Tumours
Metallic Ni powder	20	65
Crystalline Ni_3Se_2	23	91
Crystalline NiSe	16	50
Crystalline N_3S_2	9	100
Crystalline βNiS	14	100
Amorphous NiS	10	0
Vehicle Control	40	0

[1]Adapted from Sunderman FW Jr (1981)

of a preliminary report on an ongoing study (Sunderman et al 1979). The authors attributed the extreme difference between the tumour incidence induced by crystalline nickel monosulphide and amorphous nickel monosulphide to the physical form of the nickel sulphides, which they concluded to be critical to their carcinogenicity. The dissolution halftime (T_{50}) of amorphous nickel monosulphide is only 24 days in rat serum compared to 2.6 years for the crystalline nickel monosulphide (Kuehn and Sunderman 1982).

When comparing the relative solubility and tumorigenicity of various nickel compounds, we showed that nickel (II) hydroxide induced tumours in 15 of 20 rats on bilateral intramuscular injection (19 of 40 sites), while the highly soluble nickel sulphate ($NiSO_4$) gave no tumours in 27 rats from bilateral injections of 5 mg per thigh (Gilman 1965). Recently Kasprzak et al.(1983) have further investigated the carcinogenicity of nickel sulphate and several preparations of nickel hydroxide that are similar to ones encountered industrially. Table IV compares a measure of their solubilities with the tumour responses reported. These results are in general agreement with both the concept of an inverse solubility/carcinogenicity relationship and of the interdependence between the chemical structure or form of the compounds and their carcinogenicity.

CLASSIFICATION OF EXPERIMENTAL MUSCLE TUMOURS

Histological features of nickel induced rhabdomyosarcomas may range from anaplasia through every stage and admixture of pleomorphism, to true mixed tumours. The traditional

TABLE IV. $Ni(OH)_2$ and $NiSO_4$ carcinogenicity[1]

Compounds	T_{50}[2]	No. Rats	Tumours Rhab.	Tumours Other
Colloidal $Ni(OH)_2$	2-4 m	13	0	0
Air dried $Ni(OH)_2$	15 d	19	4	1
Crystalline $Ni(OH)_2$	78 d	20	3	0
$Ni(ii)SO_4$	-	20	0	0
Ni_3S_2	178 d	20	15	1
Na_2SO_4	-	20	0	0

[1]Adapted from Kasprzak et al. (1983)
[2]Dissolution time in artificial lung fluid

histopathology criterion that demands demonstrable cross-striations in a sarcoma before designating it as a rhabdomyosarcoma is unrealistic, at least in experimental tumorigenesis studies and has unquestionably led to the under-recognition of this class of sarcoma.

Yamashiro et al (1980, 1983) on the basis of light microscope studies on 70 Ni_3S_2 induced tumours, divided them into six morphological classes. Four tumours from each class were examined by electron microscopy for their ultrastructural features. Three of the six classes (A, B and C) were unequivocably subtypes of rhabdomyosarcomas, while class E was probably also myogenic. Collectively these four classes constituted 76% of the sarcomas examined.

Class A (33%) was distinguished histologically by the preponderance of pleomorphic but often well differentiated muscle elements compared to the poorly differentiated cells of Type B (27%). Similar fine structural features could be found in both, although myofilaments were much better developed and organized in Type A cells. Uninucleated myoblasts, with and without filaments, were numerous and often in mitosis. Secretory vesicles were abundant and desmosomes were sometimes seen.

Class C (7%) showed myxomatous regions interspersed with predominantly spindle cell areas typical of embryonic tissue. The ultrastructure of the spindle cells was similar to that seen in myoblasts with filaments in the pleomorphic Class A tumours.

Class E was composed of interwoven fasciculi of smooth muscle-like cells which under electron microscopy however failed to show either filaments or dark bodies, although basement membrane-like features were observed.

Friedman and Bird (1969) characterized 17 nickel induced tumours from Sprague-Dawley rats according to a system designating them all as sub-types of embryonal sarcomas (Ashton and Morgan 1965) with the following distribution:

12% Embryonal Sarcomas
18% Nonstriated embryonal rhabdomyosarcomas
70% Striated embryonal rhabdomyosarcomas

Controversy over the criteria for classification of rhabdomyosarcomas has not been resolved. This presents an ongoing problem that needs to be seriously addressed to facilitate both the histopathological diagnosis and experimental study of muscle tumours. The recently reported use of immunoperoxidase indirect sandwich technique to identify human myosin-containing rhabdomyoblasts should prove extremely helpful (Corson and Pinkus 1981).

SPECIES DIFFERENCES

Table V shows the rat and the rabbit to both be seemingly excellent models for the production of a high yield of muscle

TABLE V. Species differences (Ni_3S_2-i.m.)

Species and Dose (mg)		No. Animals (% Tumours)		Tumour Type (%) Rhab.	Other
Syrian	(5)	15	(33)	20	80
Hamster[a]	(10)	17	(71)	50	50
Mice[a]	(2.5)	20[1]	(55)	few	most
Mice[b]	(5)	45[2]	(60)	10	90
Rabbit[c]	(80)	16[3]	(100)	most	few
Rat[d]	(5)	38	(92)	54	44
Rat	(10)	23	(96)		
Rat[e]	(10)	63	(94)	66	34

[1]DBA/2 and C57Bl/6; [2]C_3H and Swiss outbred
[3]Bilateral injections; exact nos not stated
[a]Sunderman (1982); [b]Gilman (1962); Hildebrand and Biserte (1979); [d]Sunderman (1979); [e]Yamashiro et al. (1980).

tumours. The white muscle of the rabbit is of particular interest because of its capacity to give rise to both rhabdomyosarcomas and leiomyosarcomas (Hildebrand and Biserte 1979).

Limited studies on Syrian hamsters and mice have demonstrated relatively low incidences of rhabdomyosarcomas, which were generally poorly differentiated. The preponderance of hamster and particularly of mouse tumours were either fibrosarcomas or undifferentiated sarcomas. Further studies on these species are indicated, particularly at the ultrastructural level.

Another species difference worthy of note concerns frequency of metastases. This is generally very high in rats and hamsters but very low in mice, with the incidence in rabbits not reported. A metastatic lesion, particularly in the lung, will not infrequently exhibit a degree of striated muscle cell differentiation so much greater than that of its primary tumour as to necessitate revision of the initial histological classification. This suggests that in a heterogeneous tumour cell population, myoblasts programmed towards (normal) differentiation may be at a selective advantage in metastatic colonization. If this assumption is correct then it seems that ultrastructural and certain *in vitro* studies might find such lung metastases a very useful cell source.

STRAIN DIFFERENCES

Neither $C_{57}Bl$ and DBA nor C_3H and Swiss outbred strains have shown any differences in overall tumour instance, or proportions of rhabdomyosarcomas. Mice have, however, seen very limited use in these experiments and, with the great variety of inbred strains available, it would be surprising indeed if such differences were not to exist.

Extensive dose response data on rats have been published, by Sunderman (1979) on tumorigenesis with Ni_3S_2 by intramuscular(i.m.)injection. Experiments undertaken by Daniel (1966) in our laboratory showed the existence of a significant reduction in tumour incidence and extension of the latent period in NIH black rats compared to those of the Long Evans Hooded strain (Table VI). The histological tumour class distribution summarized in this table show the majority of tumours induced in the NIH black strain to have been of a mixed type, almost all with very poorly differentiated rhabdomyosarcomatous elements.

Table VII compares tumour response and rhabdomyosarcoma incidences in several commonly used rat strains. The Sprague-Dawley rats appear to be somewhat resistant to tumour induction both by Ni_3S_2 and nickel sponge. However in this instance, resistance was not expressed either in the proportions of rhabdomyosarcomas recorded (82% of 11 Ni_3S_2 sarcomas and 100% of 6 Ni sarcomas) or in their differentiation (Friedmann and Bird 1969).

TABLE VI. Strain differences; Hooded and NIH.B1 rats[1]

Strain and No.	Tumour %	Response ALP (wk)	Rhabdo. Pleo.[1]	Rhabdo. Anapl.[1]	Other Sarc.
Hooded-28	96	21	34	19	8
NIH B1-23	30	38	1	11	13

[1]Adapted from Daniel (1966). Bilateral Ni_3S_2-20 mg per thigh. Many tumours were mixed sarcomas and are listed under more than one class.

TABLE VII. Strain differences in rats to Ni_3S_2

Strain and Dose (mg)	% Tum. Sites	% Rhabdo.	% Other Sarc.
Sprague - Dawley (20)[1]	37	82	18
Hooded (10)	96	91	9
Fischer (10)	78	87	13
Wistar (10)	82	86	14

[1]Friedmann and Bird (1969)

SITE AND ORGAN SPECIFICITY

Table VIII shows that parentral injections of nickel subsulphide at other than the intramuscular site will also induce local rhabdomyosarcoma development, even in locations such as the testis that do not normally contain striated muscle fibers in the rat. Similar intratesticular rhabdomyosarcomas have also been induced with Ni_3S_2 in Syrian hamsters (Sunderman 1982). The development of striated muscle tumours in such locations supports the view that nickel exercises a selective preference for potentially myogenic cells, and also implies that retrodifferentiation of pre-existing muscle fibers need not be involved in the development of muscle tumour stem cells.

TABLE VIII. Effect of site on tumour incidence

Site and Dose (mg)	No. Rats	No. Tum.	% Rhabdo.	Reference
s.c. (20)	32	28	67	Gilman (1965)
i.p. (10)	37	9	89	" "
i.s.[2] (5)	20	4	25	" "
i.t. (10)	19	16	19	Danjanov (1978)
i.r.[2] (5-10)[1]	42	23	9	Sunderman (1979)

[1]Two dosage groups; definite classification of these tumours proved difficult
[2]i.s. Intrasplenic; i.r. Intrarenal

MECHANISMS IN MUSCLE TUMORIGENESIS BY NICKEL

(a) Ultrastructural studies

Hildebrand and Tetaert (1981) have presented convincing evidence that the origin of three Ni_3S_2 induced leiomyosarcomas was from tumour myoblasts of the striated muscle type, based on the presence in both of onco-fetal myosin and synthetic filaments with the same ultrastructure. These findings argue in favour of the tumoral stem myoblasts having been derived from redifferentiating skeletal muscle elements (retrodifferentiation concept).

The elegant ultrastructural studies of Hildebrand and Biserte (1978) and Bruni and Rust (1975) provide morphological evidence as to the origin of rat muscle tumour stem cell(s) and also lend credence to Herchen's (1963) light microscope observation that Ni_3S_2 induced rhabdomyosarcomas originated from transformed mononucleated cells (sarcoblasts) derived from the dedifferentiation of muscle fibers in the presence of nickel.

(b) Serial studies on pretumour and development stages

Surprisingly few studies using tissues from these stages of tumorigenesis have been undertaken. Herchen (1963) in our laboratory implanted solid discs of Ni_3S_2 and Fe_2O_3 into opposite thighs of Fischer rats for surgical removal at 2-4-8 to 256 day intervals. The time required for irreversible change to malignancy proved to be >32 but <64 days (Herchen and Gilman 1964).

The histology adjacent to the disc was determined for each

removal time on groups of five animals. The pattern of these changes is shown diagrammatically in Fig. 1. It was concluded (1) that the tumours, all of which were rhabdomyosarcomas, had arisen from dedifferentiation of muscle fibers induced by

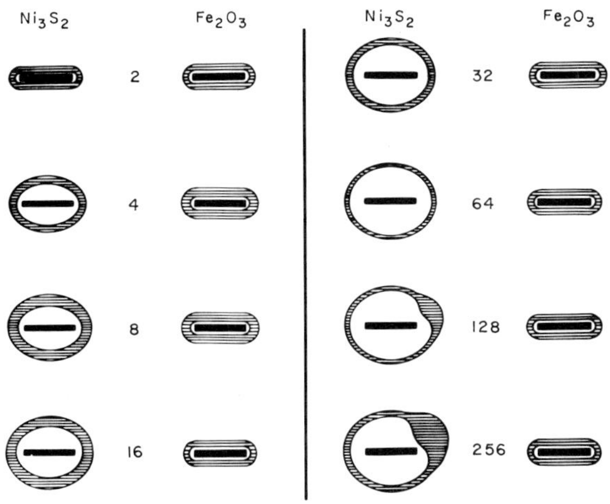

FIG. 1. Tissue changes adjacent to metal implants. Clear inner zone of necrosis. Hatched zone of muscle retrodifferentiation and early tumour formation.

a specific and critical concentraion of soluble nickel: (2) that neoplastic transformation occurred in surviving sarcoblasts (tumour stem cells) within the zone of sublethal injury; (3) that transformed sarcoblasts were resistant to nickel toxicity. A repetition and extension of these studies, in which nickel was also implanted i.m. within millipore diffusion chambers is summarized in Table IX. These experiments demonstrated conclusively that malignant transformation in the adjacent muscle tissue was chemically mediated, and seem to have precluded phagocytosis as being a necessary factor in this process (Gilman and Herchen 1963).

A further serial examination of pretumourous changes at the injection site of Ni_3S_2 has compared the tissue responses of NIH black, Fischer and Hooded rats to powdered Ni_3S_2 (Daniel 1966). Phagocytic activity was shown to be greatly enhanced,

TABLE IX. Implant form & tumorigenesis in rats

Implant Form & Dose	Incidence Tum/Rat	Sarcoma Distribution No. Exam.	% Rhab.	% Other
Ni_3S_2; 2 μ dust 10 mg	19/20	18	83	17
Ni_3S_2; 5 μ dust 10 mg	20/20	17	88	12
Ni_3S_2; chips to 500 mg	5/7	5	80	20
Ni_3S_2 discs 500 mg	14/17	14	86	14
Fe_2O_3 discs 350 mg	0/20	0	0	0
Diffusion chamber 10 mg Ni_3S_2	14/17	12	75	25
Diffusion chamber empty	1/19	1	0	100

much more persistant and to be accompanied by a massive removal of injected particles to regional lymph nodes in the resistant NIH black strain. It is suggested that the ingestion of particles by these cells reduces the Ni_3S_2 available for chemical reduction to reactive nickel, to a subcarcinogenic level.

Particular attention is drawn to these findings in view of the wide spread interest that recent observations on the phagocytosis of various nickel particles in vitro have stimulated. A causal relationship between neoplastic transformation of Chinese hamster ovary (CHO) cells and phagocytic indices in Syrian hamster embryo (SHE) cells has been shown in tissue culture systems (Costa et al. 1981). In the case of Ni_3S_2 this has been linked to in vitro carcinogenicity in nude mice (Costa et al. 1979). These and numerous similar reports on the engulfment of particles by various cell lines in tissue culture are unquestionably of very considerable interest; however, it seems timely to emphasize that extrapolations from tissue culture to the rat should be undertaken with, at the very least, just as many reservations as in those from rats to man. A recent report shows no significant rank-correlation between these phagocytic indices and the intramuscular sarcoma incidence in rats to a series of nickel compounds (Kuehn et al 1982).

RHABDOMYOSARCOMA CELLS IN CULTURE

The growth, morphology, karyotype and immunogenicity of cultured rhabdomyosarcoma cells have been the subject of several investigations. Cells from two Ni_3S_2 induced Fischer rat rhabdomyosarcomas were compared through 10 or more months of serial culture (Basrur and Gilman 1963). Primary cultures from both strains produced rhabdomyosarcomas on reinjection in rats of the same inbred strain. However, subsequent cultures of one strain (ALA) lost their tumorigenicity while those of the other became enhanced through serial passages. The loss of malignancy in the ALA strain was accompanied by fibroblast-like morphological changes and a massive shift towards the diploid chromosome number (Fig. 2).

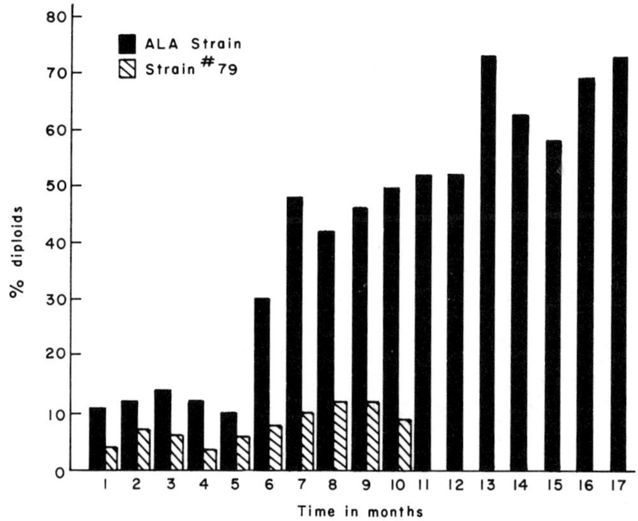

FIG.2. Per cent diploid nuclei in sequential serial cultures from two nickel subsulphide-induced rat rhabdomyosarcomas.

Recent investigations by Pot-Debrun et al. (1983) have used a series of cloned cell lines from a primary Ni_3S_2 induced rat rhabdomyosarcoma in studies that reemphasize the heterogeneity of tumour cell populations. The use of this model would seem particularly useful in identifying the metastatic and immunological differences between tumour cell populations such as that of the inverse relationship to spontaneous and experimental metastases (Sweeney et al 1982).

The morphological and synthetic responses of normal neonatal rat muscle and nickel induced rhabdomyosarcoma cells in culture to Ni_3S_2 have also shown the tumour cells to be heterogeneous, to have a limited capacity to differentiate, and to be more resistant to the cytotoxic effects of the nickel in the media (Basrur and Gilman 1967; Corbeil 1967). Nickel was also shown to suppress the mitotic index and induce aberrant mitoses through interference in spindle fiber formation and contractile protein synthesis (Swierenga and Basrur 1968).

CONCLUSIONS

Solubility and the physical form of the test chemical appear to be major factors influencing the neoplastic activity of nickel compounds. The affinity of nickel for muscle was not appreciably affected by any of the variables discussed. This affinity has provided a consistently useful model for studies on the mechanisms involved in the genesis of sarcomas and particularly on muscle tumour development.

Probably as a reflection of the current emphasis on genotoxicology, interest in in vitro studies on various nickel muscle interactions again seems to be increasing. These should prove rewarding if pursued with imagination using the sophisticated labeling techniques now available.

REFERENCES

Ashton N. and Morgan G. (1965) Journal of Clinical Pathology, 18, 699-714.

Basrur, P.K. and Gilman, J.P.W. (1963) Journal of National Cancer Institute, 80, 163-201.

Basrur, P.K. and Gilman, J.P.W. (1967) Cancer Research, 27 (1), 1168-1177.

Bruni, C. and Rust, J.N. (1975) Journal of National Cancer Institute, 54, 687-696.

Corbeil L.B. (1967) Cancer 20, 572-578.

Corson, J.M. and Pinkus, G.S. (1981) American Journal of Pathology 103, 384-389.

Costa M., Nye J.S., Sunderman F.W. Jr., Allpass P.R. and Gondos B. (1979) Cancer Research 39, 3591-3597.

Costa M., Abbracchio M.P. and Simons-Hansen J. (1981) Toxicology and Applied Pharmacology 60, 313-323.

Costa M., Heck J.D. and Robinson S.H. (1982) Cancer Research 42, 2757-2763.

Daniel M.R. (1966) The British Journal of Cancer 20, 886-895.

Danjanov I., Sunderman F.W. Jr., Mitchell J.M. and Allpass P.R. (1978) Cancer Research 38, 268-276.
Eckhardt R.E. (1959) Industrial Carcinogenesis, pp 98-99. Grune and Stratton, New York.
Friedmann I. and Bird E.S. (1968) Journal of Pathology, 97, 375-382.
Gilman J.P.W. (1962) Cancer Research 22, 158-162.
Gilman J.P.W. and Ruckerbauer G. (1962), Cancer Research, 22, 152-157.
Gilman J.P.W. and Herchen H. (1963) Acta Union International Contra le Cancer 19, 615-619.
Gilman J.P.W. (1965) Proceedings of the Sixth Canadian Cancer Research Conference, pp 209-223. Pergamon Press, Oxford.
Herchen H. (1963) A Study of the Genesis of Rhabdomyosarcomas Induced with Nickel Sulfide in the Rat. MVSc Thesis, University of Toronto, Canada.
Herchen H. and Gilman J.P.W. (1964) Nature 202, 306-307.
Hildebrand H.F. and Biserte G. (1978) Cancer 42, 528-554.
Hildebrand H.F. and Biserte G. (1979) Cancer 43, 1358-1374.
Hildebrand H.F. and Tetaert D. (1981) Onco-developmental Biology and Medicine 2, 101-108.
Hueper W.C. (1952) Texas Report on Biology and Medicine 10, 167-186.
Hueper W.C. (1955) Journal of the National Cancer Institute 16, 55-73.
International Agency for Research on Cancer (1976). Monograph on the Evaluation of Carcinogenic Risk of Chemicals to Man, vol 11, pp 75-112. IARC, Lyon, France.
Kasprzak K.S., Gabryel P. and Jarczewska K. (1983) Carcinogenesis, 4, 275-279.
Kuehn K. and Sunderman F.W. Jr. (1982) Journal of Inorganic Biochemistry 17, 29-39.
Kuehn K., Fraser C.B. and Sunderman F.W. Jr. (1982) Carcinogenesis 3, 321-326.
Moon Ko-Kim (1969) Pulmonary Effects of Metallic Dusts. Nickel and Iron. MSc Thesis, School of Hygiene, University of Toronto, Canada.
Payne W.W. (1964) Proceedings of the 55th Annual Meeting of the American Association for Cancer Research (Abst. #197).
Pot-Debrun J., Poupon M-F., Sweeney F.L. and Chouroulinkov I. (1983) Journal of the National Cancer Institute 71, 1241-1245.
Sunderman F.W. Jr., Taubman S.B. and Allpass P.R. (1979a) Annals of Clinical and Laboratory Science 9, 441.
Sunderman F.W. Jr. (1979b) in Emnelot P. and Krick E. (Eds), Environmental Carcinogenesis pp 165-192. Elsevier/North Press, Amsterdam.

Sunderman F.W. Jr. (1981) Environmental Health Perspectives 40, 131-141.
Sunderman F.W. Jr. (1982) in Langenbach R., Nesnow S. and Rice J.W. (Eds), Organ and Species Specificity in Chemical Carcinogenesis pp 107-127. Plenum Press, New York.
Sweeney F.L., Pot-Debrun J., Poupon M-F. and Chouroulinkov I. (1982) Cancer Research 42, 3776-3782.
Swierenga S.H.H. and Basrur P.K. (1968) Laboratory Investigation 10, 663-674.
Yamashiro S., Gilman J.P.W., Hulland T. and Abandowitz H.M. (1980) Acta Pathologica Japonica 30, 9-22.
Yamashiro S., Basrur P.K., Gilman J.P.W., Hulland T. and Fujimoto Y. (1983) Acta Pathologica Japonica 33, 45-58.

CARCINOGENESIS

IN VITRO STUDY OF Ni_3S_2-INDUCED RHABDOMYOSARCOMAS IN NORMAL AND POLYMYOPATHIC HAMSTERS.

GAETAN JASMIN and CONSTANTIN TAUTU

Department of Pathology, Faculté de médecine, Université de Montréal, Montreal, Canada H3C 3J7.

4.

INTRODUCTION

Recent studies on the carcinogenicity of Ni_3S_2 in polymyopathic hamsters indicate that in spite of a defective myogenesis (Tautu and Jasmin 1980; Tautu and Jasmin 1982), these animals developed rhabdomyosarcomas at the same rate (8-12 mo) and with the same frequency (70%) as normal controls. The cytology of the primary tumors was comparable and remained constant after 10 successive *in vivo* passages. In most cases, the fibrillar nature of the malignant rhabdomyoblasts was histologically evident (Fig. 1*A*) but early sarcomeric arrangement was more easily demonstrable in EM (Fig. 1*B*). When grown *in vitro*, using a previously described explant technique (Tautu and Jasmin 1982), two categories of neoplastic cells were distinguishable by morphology and grouping. One variety was larger, less profuse and strongly acidophilic. These huge cells were often pleomorphic, displaying one or several irregularly shaped nuclei with an heterogenous chromatin and prominent nucleoli (Fig. 2*A*). Their cytoplasm, according to the state of differentiation, was granular or fibrillar with processes giving the cell a strap or spider-like appearance (Fig. 2*B*). The second variety of neoplastic cells was apparently more incipient. They were spindle-shaped, enclosing an oval nucleus of variable size within a scant lightly stainable cytoplasm (Fig. 2*C*). Following successive passages, these small cells tend to proliferate more rapidly with a shorter doubling time and loss of contact inhibition when the cultures became confluent. Myogenic foci frequently occurred in these cultures. Yet, it remains to be seen if we are frankly dealing with neoplastic cells. Thus, the question arose as whether these two types of neoplastic cells were of the same lineage, merely expressing different stages of myogenic dysmaturation.

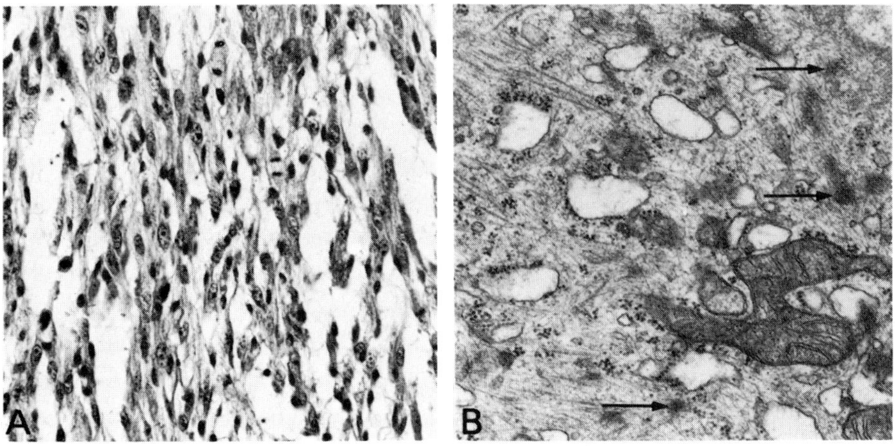

FIG. 1. Microscopic appearance of transplanted tumor cells (10th passage). *A*. The myofibrillar aspect of neoplastic cells is well preserved. X250. *B*. In EM, the myofilaments and Z band material (arrows) are conspicuous. X30,000.

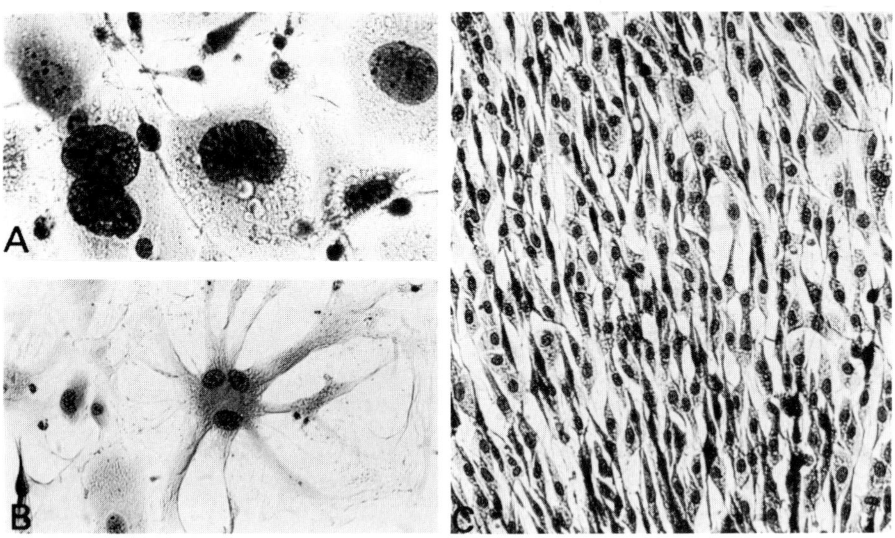

FIG. 2. Microscopic appearance of two types of neoplastic cells observed in the cultivated rhabdomyosarcomas. *A* and *B*. Large mono- or plurinucleated pleomorphic cells in primary cultures. X1,200 and X750. *C*. Aggregated spindle-shaped small cells as seen in secondary cultures. X120.

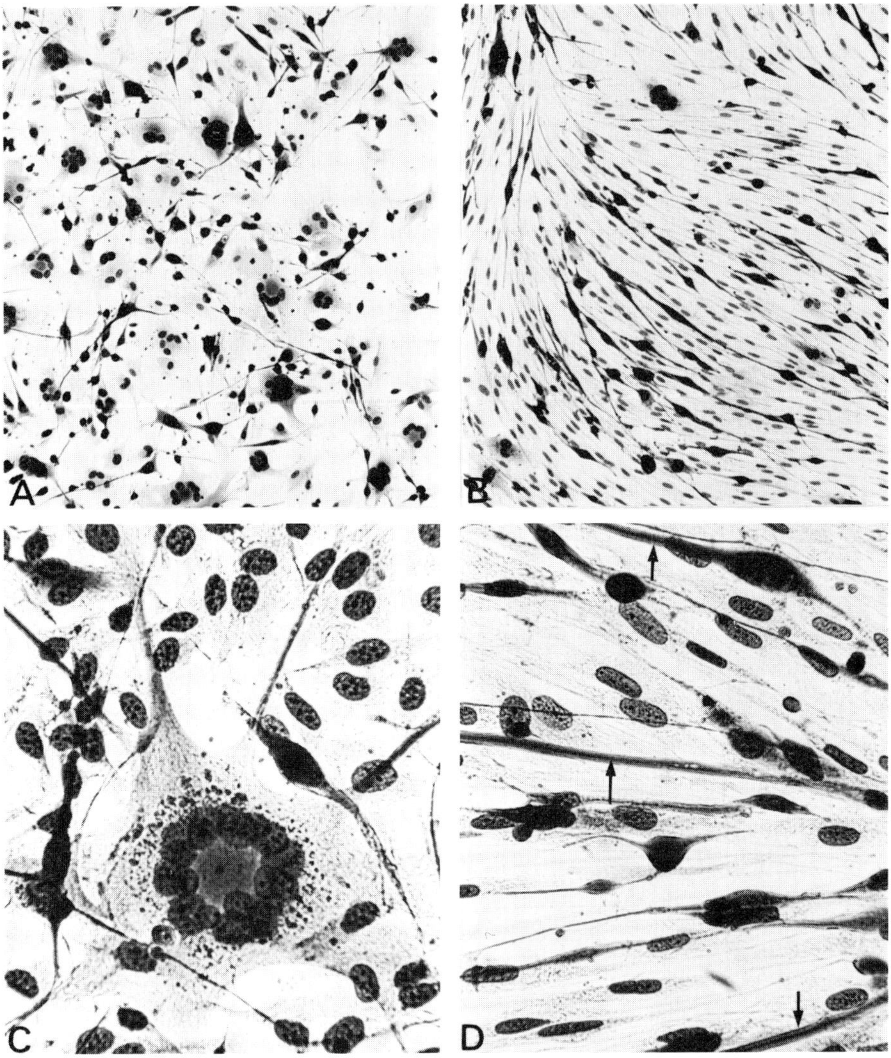

FIG. 3. Cytology of cultivated neoplastic cells after DMF treatment. *A.* After two weeks, the two types of neoplastic cells observed in primary cultures are sparsely distributed. X300. *B.* After five weeks, the tumor cells become elongated. X120. *C.* High magnification of a giant rhabdomyoblast with concentrically arranged segmented nuclei. X1,200. *D.* Fibrillar and faintly striated cytoplasmic processes (arrows) outgrowing from the giant rhabdomyoblasts. X1,200.

METHODS AND RESULTS

To investigate this question, we used N,N-dimethylformamide (DMF) which has been claimed by Dexter (1977) as being a differentiation inducer. DMF was added to the basal culture medium in a final concentration of 1%, the medium being changed twice weekly. Soon after a two-week treatment, the saturation density of these cultures was markedly depressed in comparison to untreated cultures (Fig. 3*A*). Upon the 40th day, the neoplastic cells became elongated, loosing their tendency of piling up (Fig. 3*B*). But most striking was the remodeling of the larger rhabdomyoblasts. Their giant nucleus underwent segmentation, giving rise to a collection of isomorph nuclei disposed in a doughnut fashion (Fig. 3*C*). As a result numerous deriving cells expanded in a linear pattern with long fibrillar and faintly striated processes (Fig. 3*D*); these newly-formed elongated cells admixed with other swirling small tumor cells. We further observed that the transplantability of intraperitoneally injected tumor cells was almost completely prevented by a 60-day *in vitro* treatment with DMF. Ultrastructural and chromosomal studies are in progress to further evaluate the differentiation state of these neoplastic treated cells.

SUMMARY

Malignant muscle tumors were induced by intramuscular injection of Ni_3S_2 in normal and myopathic hamsters. These tumors exhibited all features of a pleomorphic rhabdomyosarcoma. When grown *in vitro*, two categories of cells were demonstrable; their morphology and growth rate were influenced by a sustained treatment with dimethylformamide. It is inferred that these tumor cells retain some differentiation potentialities highly compatible with a progressive loss of malignity.

ACKNOWLEDGEMENTS

This work was supported by the MDA of Canada, the MRC of Canada and the Macdonald-Stewart Foundation.

REFERENCES

Dexter D.L. (1977) Cancer Research, 37, 3136-3140.

Tautu C. and Jasmin G. (1980) Journal of Neuropathology and Experimental Neurology, 39, 173-180.

Tautu C. and Jasmin G. (1982) Journal of Neuropathology and Experimental Neurology, 41, 87-103.

EFFECTS OF CALCIUM AND MAGNESIUM SALTS ON NICKEL SUBSULFIDE CARCINOGENESIS IN FISCHER RATS

KAZIMIERZ S. KASPRZAK and LIONEL A. POIRIER
Nutrition and Metabolism Section,
Laboratory of Comparative Carcinogenesis,
Division of Cancer Etiology, National Cancer Institute,
Frederick, MD 21701, USA

INTRODUCTION

The physiologically essential divalent metals zinc and manganese have long been known to antagonize the carcinogenicity of cadmium (Parizek, 1957) and nickel (Sunderman et al., 1974), respectively. More recently, it has been found in our laboratory that the two major essential divalent metals, magnesium and calcium, also prevent tumor induction by other metals. Magnesium prevented the induction of local tumors by cadmium in rats (Poirier et al., 1983), and both magnesium and calcium prevented the tumorigenic activities of nickel and lead in the lungs of strain A mice (Poirier et al., 1984). The mechanisms underlying such protection are not clear and more experiments must be completed before any generalizations are made as to the nature and practical importance of the observed phenomena. In the present study we examined the ability of calcium and magnesium to inhibit the carcinogenic activity of nickel subsulfide in the rat muscle. Calcium and magnesium salts were administered by three different routes: in the diet, by distal subcutaneous injections, or by local injection in a mixture with the carcinogen.

EXPERIMENTAL

Eleven groups of male weanling Fischer rats, 20 rats/group, received a single injection of 2.5 mg of nickel subsulfide (Ni_3S_2) into both thigh muscles (i.m.). The rats were then treated as follows. Four groups were fed 3% calcium acetate (CaAcet) or 3% magnesium acetate (MgAcet) diets for 3 or 6 months. The next four groups of rats were given subcutaneous (s.c.) injections of 0.16 mmol CaAcet/kg/inj., or 2 mmol MgAcet/kg/inj. in the thoracic region, 3 times weekly for 1 or 4 months. The corresponding 3 control groups of rats were fed basal Purina diet and received s.c. injections of 0.15 M saline or 4 mmol sodium acetate (NaAcet)/kg/inj., 3 times weekly for 4 months.

Two separate groups of male weanling Fischer rats, 20 rats per group, were given i.m. injections of 2.5 mg Ni_3S_2 mixed with 6.2 mg of anhydrous calcium carbonate (CaCarb) or with 6.3 mg of magnesium basic carbonate ($4MgCO_3 \cdot Mg(OH)_2 \cdot \underline{n}H_2O$; 40.6% MgO; MgCarb) in a 0.15 M saline suspension. The corresponding 3 control groups of rats received i.m. injections of saline, 6.2 mg CaCarb, or 6.3 mg MgCarb, respectively.

All rats were observed for 18 months. The tumors were examined histologically. Statistical analysis of the results was performed according to Thomas et al. (1977).

The amounts of nickel, calcium, and magnesium remaining at the injection sites at 5 hr to 16 wk after the injections were determined by atomic absorption in separate groups of rats, 3 rats per time point, under the same conditions as in the carcinogenicity trials.

RESULTS

No acute toxic effects of any of the treatments were observed. Only one rat died without tumors before the sacrifice.

The cumulative muscle tumor incidence curves are shown in Fig.1. The final yield of the injection site tumors after 18 months of experiment equalled: 1) 70-90% of the rats injected with Ni_3S_2 and fed basal diet or the diets containing CaAcet, or MgAcet, with no significant differences among the groups; 2) 95-100% of the rats given i.m. Ni_3S_2 followed by multiple s.c. injections of CaAcet, MgAcet, NaAcet, or saline, with no significant differences among the groups; and 3) 85% of the rats injected with Ni_3S_2 + CaCarb, 25% of the rats injected with Ni_3S_2 + MgCarb, and 0% of the rats injected with CaCarb, MgCarb, or saline alone.

The muscle tumors were predominantly unilateral rhabdomyosarcomas and fibrosarcomas. No significant differences were noticed in the distribution of various histological types of sarcomas among the groups. Also, the generally higher tumor yields in rats treated with the multiple s.c. injections were not statistically different from the tumor yields in other treatment groups.

The results of determinations of nickel, calcium, and magnesium in the injected muscles are shown in Fig.2. As can be seen in this Figure the admixtures of either CaCarb or MgCarb to Ni_3S_2 did not exert any significant influence on the mobilization of nickel from the injected muscle. Magnesium was the fastest mobilized metal of the three tested; 98% of its dose was removed in one week, compared to 70% of the calcium dose, and 20% of the nickel dose.

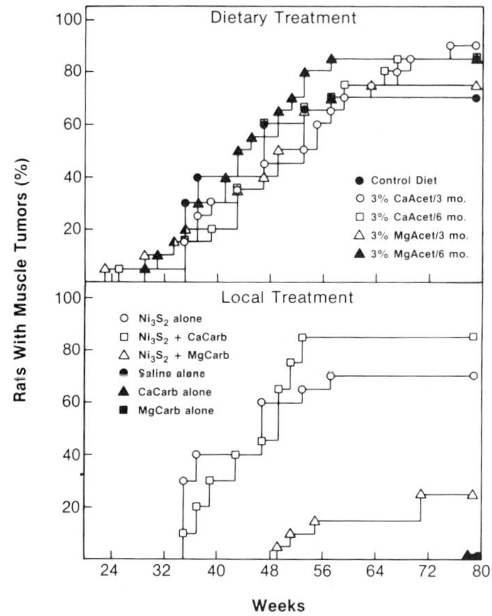

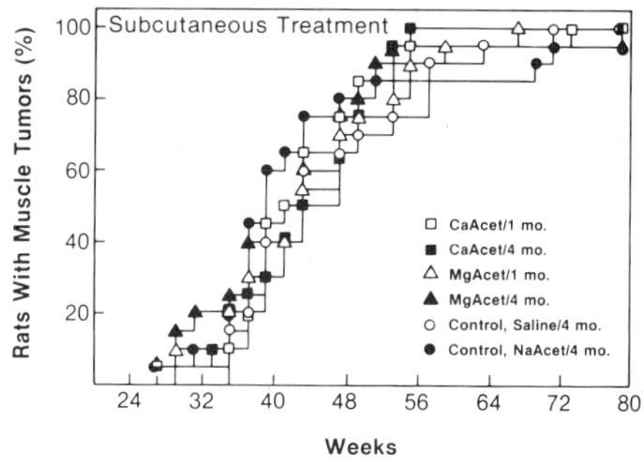

FIG.1. Cumulative incidences of muscle tumors in rats injected i.m. with nickel subsulfide and treated with calcium or magnesium salts.

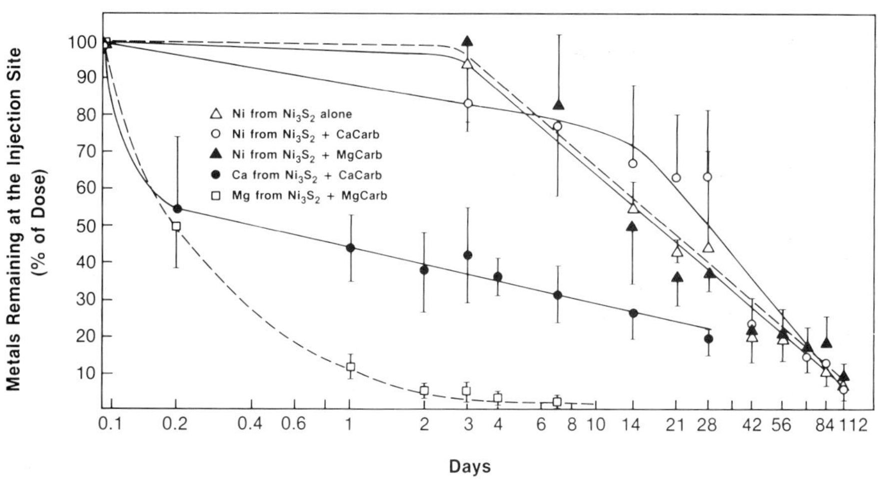

FIG.2. Fractions of the metal doses remaining at different times at the site of injection. Each point represents the mean value for 3 rats (6 injection sites) $\pm$ S.E.M.

DISCUSSION

The results of the present study demonstrate that the induction of muscle tumors by nickel subsulfide is markedly inhibited by a simultaneous local administration of magnesium basic carbonate. Calcium carbonate under the same experimental conditions does not produce such inhibition. Likewise, the treatment with calcium or magnesium acetate, by either dietary or distal s.c. route, does not result in any inhibition of the nickel subsulfide carcinogenicity. Thus, to exert its protective effect magnesium must be present at the site of nickel action in relatively high concentrations. This cannot be achieved by a prolonged oral or s.c. administration of magnesium, even at doses close to the maximum tolerated doses (Poirier et al., 1983), as used in this experiment. Another important finding of this study is the fast mobilization of magnesium from its injection site. Hence, the elevated concentration of the protective metal is maintained only during a very short period relative to the latency of the nickel carcinogenesis. This might indicate that some critical events, possibly responsible for initiation, occur in muscles injected with nickel subsulfide in the few days following the injection.

CONCLUSION

The present results do not permit any definite conclusions to be drawn as to the nature of the observed anticarcinogenic effects of magnesium. Perhaps these effects may result from similarities in the physicochemical properties of the Mg^{2+} and Ni^{2+} ions (Nieboer and Richardson, 1980) that make them able to compete for common binding biomolecules. If so, the carcinogenic activity of nickel might occur through a displacement of magnesium from its cellular functional sites.

REFERENCES

Nieboer E. and Richardson D.H.S. (1980) Environ. Pollut., Ser.B, 1, 3-26.

Parizek J. (1957) J. Endocrinol., 15, 56-63.

Poirier L.A., Kasprzak K.S., Hoover K.L. and Wenk M.L. (1983) Cancer Res., 43, 4575-4581.

Poirier L.A., Theiss J.C., Arnold L.J. and Shimkin M.B. (1984) Cancer Res., 44, 1520-1522.

Sunderman F.W., Jr., Lau T.J. and Cralley L.J. (1974) Cancer Res., 34, 92-95.

Thomas D.G., Breslow D. and Gart J.J. (1977) Comput. Biomed. Res., 10, 373-381.

NICKEL-INDUCED MALIGNANT TUMORS

GEORGE LUMB*, F. WILLIAM SUNDERMAN SR.*†,
and HENRY P. SCHNEIDER†
*Department of Pathology,
†Institute for Clinical Science,
Hahnemann University,
Philadelphia, Pennsylvania 19102 USA

Nickel-induced malignant tumors may be produced in experimental animals by the inhalation of nickel carbonyl and by implantation of nickel compounds into various tissues. The present studies have been undertaken on the pathogenesis of nickel-induced tumors by the implantation of nickel subsulfide encased in small gelatin capsules into the hind limb muscles of Fischer 344 rats.

The antitumorigenic effectiveness of sodium diethyldithiocarbamate was studied in rats receiving nickel subsulfide implantations. These studies reveal a striking sex responsiveness to treatment. Only 12 percent of treated female rats developed tumors as compared to 52 percent of treated males. In our series, 78 percent of untreated rats developed tumors. Castration of male rats did not produce any significant change in the responsiveness to therapy.

Histomorphologic studies have been made on 78 tumors, 13 metastases, and one transplanted tumor. In addition to the usual paraffin embedding, plastic embedding was used to procure ultra thin sections of one micrometer thickness. Plastic embedding extends the restricted observations of electron microscopy (which was also used) by allowing larger samples of tissues to be studied.

The cells found in the tumor include fibroblast-type and histiocyte-type with numerous bizzare giant cells. A storiform (cart wheel) pattern was seen in both primary and metastatic tumors (Fig.1).

Electron microscopy revealed spindle shaped cells with abundant rough endoplasmic reticulum (RER) and collagen formation in intercellular areas. These were fibroblast-like cells (Fig.2). Ovoid cells with microvilli and well-defined Golgi but less abundant RER and no collagen formation were designated histiocyte-like. Intermediate cells with some features of both types were seen in small numbers. Giant cells were confirmed. No rhabdomyoblasts were seen (Fig.3).

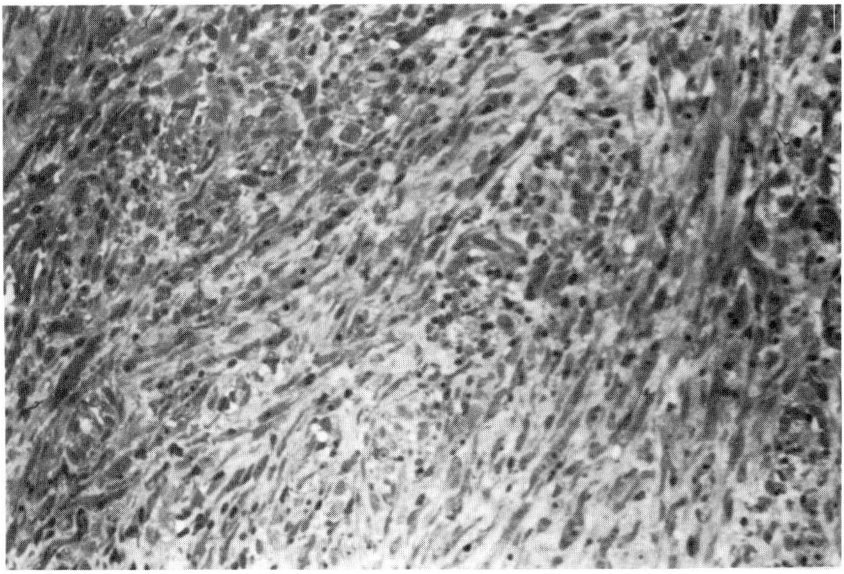

FIG.1. Mostly spindle cells arranged in a "storiform pattern. H & E x 100.

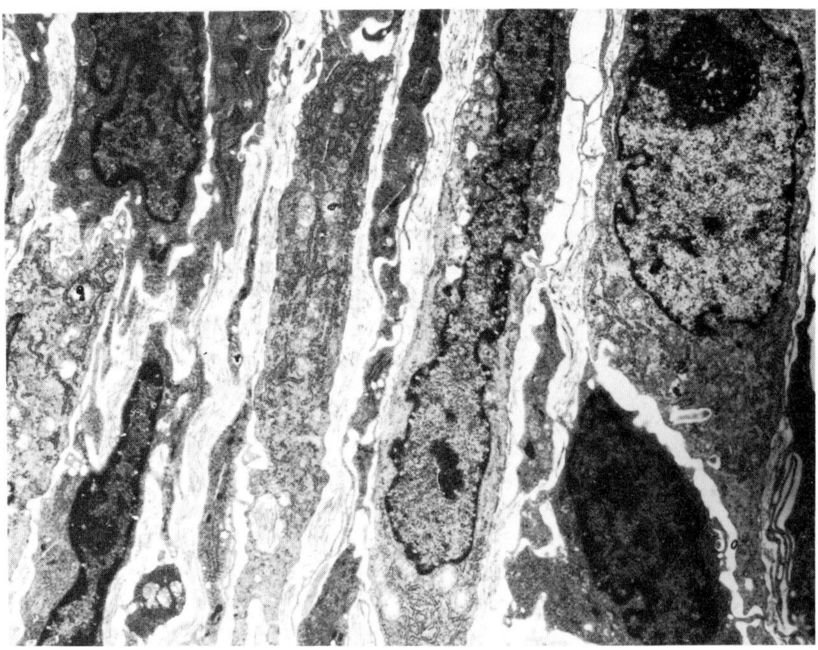

FIG.2. Electron micrograph of fibroblast-type and histiocyte-type cells. x 800

These appearances indicate a diagnosis of malignant fibrous histiocytoma (MFH) as being the most appropriate for the experimentally induced tumors in this series.

Malignant fibrous histiocytoma is the most frequent soft tissue sarcoma in humans of older age group, and its origin from pluripotential mesenchymal cells is considered likely. Such a theory would be attractive as the source of origin of the experimental tumors. It would account for the variety of tumor types occurring following Ni_3S_2 implantation into muscle. Although rhabdomyosarcoma has been the most frequently described tumor following intramuscular Ni_3S_2 implantation, it is interesting to note that 48 percent of Sunderman's (1983) intramuscular tumor series were designated soft tissue sarcoma other than rhabdomyosarcoma.

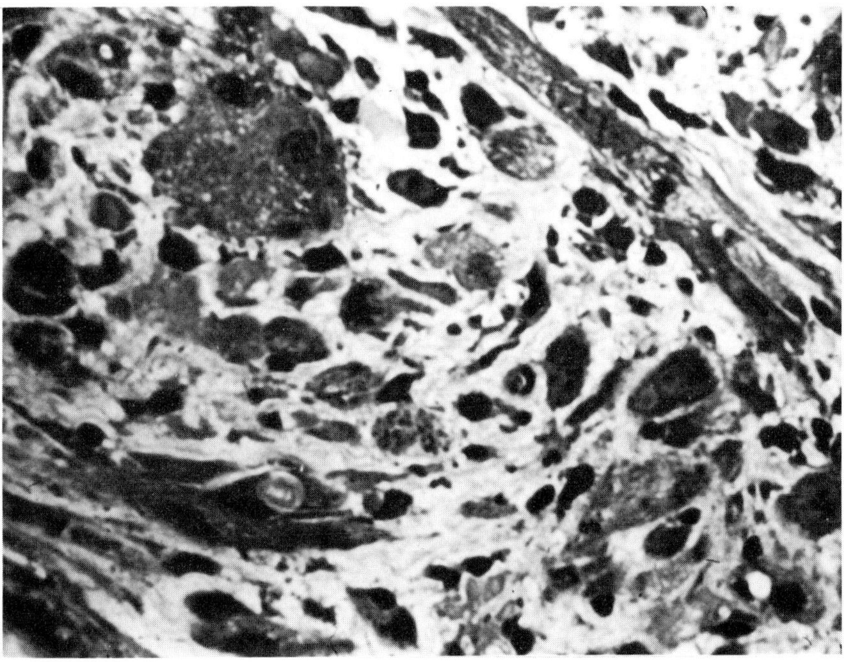

FIG.3. Mixed cell population of spindle, ovoid, and giant cells. One micron section. Toluidine blue x 400.

REFERENCE

Sunderman F.W. Jr. (1983) in Langenbach R., Nesnow S. and Rice J. (Eds.), Organ and Species Specificity in Chemical Carcinogenesis, pp. 107-126. Plenum Publishing Corp., New York.

TUMOR INDUCTION IN RATS AFTER INTRAPLEURAL INJECTION OF NICKEL SUBSULPHIDE AND NICKEL OXIDE

VIDAR SKAUG, BJØRN GYLSETH, ANE-LENE PALMER REISS and TOR NORSETH
Institute of Occupational Health, P.O. Box 8149 Dep, Oslo 1, Norway

INTRODUCTION

Intrapleural injection has not previously been applied for testing nickel tumorigenesis. As the mechanisms and relative potency of different nickel compounds to cause cancer are still unclear,all models which might shed light on these problems deserve testing. Intrapleural injection also allows comparism of nickel and another potent carcinogen - asbestos.

MATERIALS AND METHODS

Four groups each consisting of 32 male Wistar rats (3 months/ 200 g) were used. Each rat in the respective group was given a single intrapleural injection in 0.4 ml of saline of 10 mg nickel oxide (NiO Merck PA - Joint Committee on Powder Diffraction Standard file no. 4-835), 10 mg nickel subsulphide (Ni_3S_2 - gift from Inco Limited), 10 mg crocidolite (UICC standard), or 0.4 ml of saline alone. Crocidolite was included as a positive control in the experiment, saline served as the negative control. Autopsy was performed on all animals either shortly after death or after they had been sacrificed due to severe suffering.

RESULTS

Table 1 shows that almost all rats in the nickel groups had developed malignant tumors 30 months after the injection. Less than 50% of the animals in the crocidolite group had tumors. In the saline group only 5 animals with tumors were found. Subcutaneous fibromas found in a number of animals in all groups are not included. The cause of death for animals in the saline group was chronic nephritis, some cases were also found in other groups.

TABLE 1. Malignant tumors in rats 30 months after injection of nickel subsulphide or nickel oxide

Group	Rats dead/ total number	Rats alive	Rats with tumor
Ni_3S_2	31/32*	1	28
NiO	30/32	1	31
Croci-dolite	23/32	9	9
Saline	25/32	7	5

* Three rats killed without tumor because of apparent suffering

The first tumors in the nickel groups appeared already after 120 days. Figure 1 further shows that most of the rats in the nickel groups had developed tumors within one year.

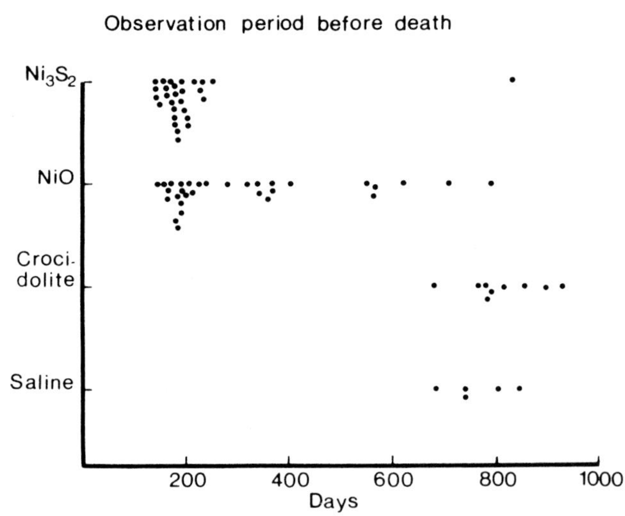

FIG. 1. Time interval between injection and death (or sacrifice) for rats with tumors. Each dot represents one animal.

The time between injection and death for the nickel subsulphide group was significantly shorter than for the oxide group, median value 177 and 224, respectively ($p<0.01$ Mann-Whitney). Comparison of the two groups also demonstrates a skewed distribution of the time intervals for the oxide group. For the crocidolite group the first tumors appeared after about 700 days. Median time interval 814 days. Tumors in the saline group appeared at about the same time. There are still after 30 months a number of rats alive in these groups.

All tumors in the nickel groups and in the crocidolite group were sarcomas. There were, however, definite histological differences between the groups. All tumors in the nickel groups except for 3 showed malignant muscle cell differentiation (rhabdomyosarcoma) while such differentiation was only present in one of the crocidolite exposed rats. Sarcomas were not demonstrated in the saline group.

There were no obvious differences between the macroscopic localization of the tumors in the nickel groups and in the crocidolite group. Invasive growth was demonstrated in all three groups as well as involvement of local lymph nodes. Aggregates of nickel particles surrounded by local inflammatory reaction and/or tumor tissue could be seen both in the muscle tissue and in what appeared to have been the intrapleural space.

DISCUSSION AND CONCLUSIONS

The tumorigenic effect of nickel particles (dust) depends on the biological availability of nickel from the particle in question. Nickel subsulphide is more soluble than nickel oxide in biological fluids, as reflected in the shorter and more uniform time period between injection and death in this group compared to the oxide group. The total dose injected seems to be of less importance as the amount of nickel injected is slightly higher in the oxide group than in the subsulphide group.

The time between injection and tumor development is significantly longer for the crocidolite group than for either of the nickel groups. The origin of the tumor also seems different. Sarcomas or mesotheliomas after asbestos injection in the pleural cavity originate from sub-mesothelial mesenchymal cells of limited differentiation. Conservatively, we have classified the tumors in the crocidolite group as sarcomas although mixed epithelial and mesenchymal components as required for the definite diagnosis of mesothelioma were also seen.

Solid state tumorigenesis has been suggested both for asbestos and slightly soluble salts for injection site tumor development. Obviously more than one mechanism is in effect when nickel and asbestos are compared. We must assume that fibers as well as nickel particles are deposited in the injection channel on withdrawal of the needle. The mechanism for fiber tumorigenesis is not known, fiber morphology seems important as well as durability in biological fluids. The fiber probably has to penetrate the cell to induce intra-cellular biochemical alterations. These factors compared to the penetration of nickel ions may thus be important for the differences in tumorigenicity which is observed. The differences also make unspecific solid state tumorigenesis unlikely.

Spindle cell sarcoma and mesothelioma have been demonstrated after subcutaneous injection of asbestos (Roe et al., 1967). There was, however, a specific migration of fibers from the injection site to the submesothelial tissues of thorax and abdomen. A more extensive local reaction was found at these sites compared to at the injection site.

Aggregates of nickel particles were demonstrated in the pleural space, and nickel uptake in the submesentherial cells must be assumed. Seemingly these cells have a lower tumorigenic potential upon nickel exposure than muscle cells. Rhabdomyosarcomas have been demonstrated after nickel injection into rats testis, an organ devoid of striated muscle. Fibrosarcoma was, however, the predominant form present (Sunderman, 1983). The origin of the tumors was probably immature mesenchymal precursor cells (Damjanov et al., 1977). We assume that testicular mesenchymal cells have a more extensive differentiation potential than the subpleural cells and that the sarcomas in our case originate from striated muscle. Microscopically we have demonstrated intermediate structures between muscle and tumor, supporting this assumption.

REFERENCES

Damjanov I., Mitchell J.M., Allpass P.R., Bigazzi P. and Sunderman F.J. Jr. (1977) Proceedings of the American Association for Cancer Research, 18, 52.

Roe F.J.C., Carter R.L., Walters M.A. and Harrington J.S. (1967) International Journal of Cancer, 2, 628-638.

Sunderman F.W. Jr. (1983) in Langenbach R., Newnow S. and Rice J. (Eds.), Organ and Species Specificity in Chemical Carcinogenesis, pp. 107-126. Plenum Publishing Corp., New York.

ELECTRON MICROSCOPY OF PULMONARY LESIONS, INCLUDING CARCINOMA, INDUCED BY INHALATION EXPOSURE OF RATS TO NICKEL OXIDE AEROSOL

AKIO HORIE, ISAMU TANAKA, JOJI HARATAKE, YASUSHI KODAMA and KENZABURO TSUCHIYA
School of Medicine,
University of Occupational and Environmental Health,
1-1 Iseigaoka, Yahatanishiku, Kitakyushu 807, Japan

INTRODUCTION

These experiments were designed to test the carcinogenic effects of nickel oxide (NiO) inhalation on the lung and this report describes the histopathological findings.

MATERIALS

NiO particles, distributed by Soekawa Chem. Industry, Japan, were reduced by a ball mill and fed to a dust generator. The median aerodynamic and the geometric standard deviation were 1.2 µm and 2.2 µm respectively.

All the animals used were male Wistar rats, about 2 months old, approximately 250g in body weight, supplied by the Animal Center of the School. They were kept in exposure chambers (0.4 m^3), five to a cage and provided with commercial pellet food.

METHOD

a) The inhalation system

The details of this system were reported by Tanaka et al. (1982). Five rats in each experimental group in exposure chamber were forced to inhale NiO particles 6 hours per day, 5 days per week for one month.

b) The experimental condition

The first group was used to confirm the inhalation system, tolerance of animals and distribution of NiO particles in the lung. The other conditions are listed in Table 1.

c) Histopathological procedures

The first and fourth experimental groups were sacrificed by

Table 1 Experimental conditions

Experimental No	1		2			3			4		
Exposure level	High	Cont	High	Low	Cont	High	Low	Cont	High	Low	Cont
NiO concentration (mg/m^3)	6.5 $\pm$3.5		7.0 $\pm$3.9	1.4 $\pm$0.5		8.0 $\pm$1.8	0.6 $\pm$0.3		8.0 $\pm$1.8	0.6 $\pm$0.3	
Exposure period	1		1	1		1	1		1	1	
Total exposure time (hour)	216		147	147		140	140		140	140	
Number of animals	2	2	5	5	5	5	5	4	8	6	5
Time sacrificed	Immediately after one month exposure					12 months after one month exposure			20 months after one month exposure		

decapitation, while the 2nd and 3rd groups were autopsied after anesthesia with sodium pentobarbital. The lungs, liver, kidney, pancreas, spleen and cervical lymph nodes were fixed in 10% formalin for light microscopy. Three-dimensional investigation by serial sections was used for the detection of NiO particles distributed in the lung. In the 4th group, frontal sections of the head were used for investigation of the upper respiratory lesions.

For electron microscopy, 1 mm cubes of the same tissues were fixed in 2.5% phosphate-buffered glutaraldehyde overnight, postfixed in 1% phosphate-buffered osmium tetroxide, dehydrated in graded ethanol series and embedded in Epon 812. JEM 10OC electron microscope was used for the investigation of the ultrastructure, and Hitachi H500 analytical electron microscopy for detection of nickel localization.

RESULTS

Analytical electron microscopy disclosed that the tiny particles contained nickel metal by image scanning. NiO particles about 1 μm in diameter were electron-dense, angular in shape and mainly trapped by the alveolar macrophages. Tubular myelin, consistent with surfactant, was noted in the alveoli of the inhaled lungs. Type II pneumocytes disclosed secretory activity of tubular myelin. Cytoplasm of Clara cells also showed prominent secretory activity and projection into the bronchiolar lumen.

In the 3rd group sacrificed at an interval of one year, serial sections of the lungs with conglomerate nodule disclosed that NiO particles were distributed in the peribronchiolar, perivascular and subpleural alveoli, hilar lymphoid apparatus and within the pockets of terminal bronchioli. Analytical electron microscopy showed some nickel particles within the lysosomes of macrophages. Pools of tubular myelin were noted in the peribronchial lymphatics. The Clara

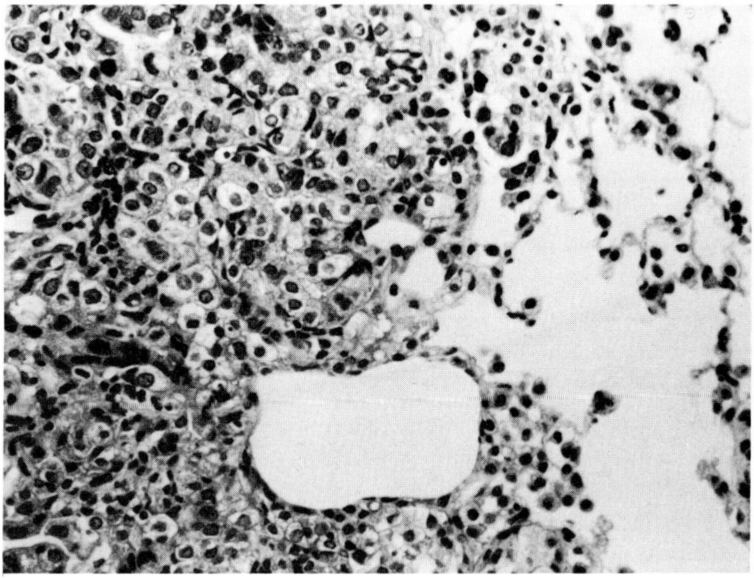

FIG. 1. Adenocarcinoma of the lung in a low-exposure-rat of the 4th group. H.E. X 120.

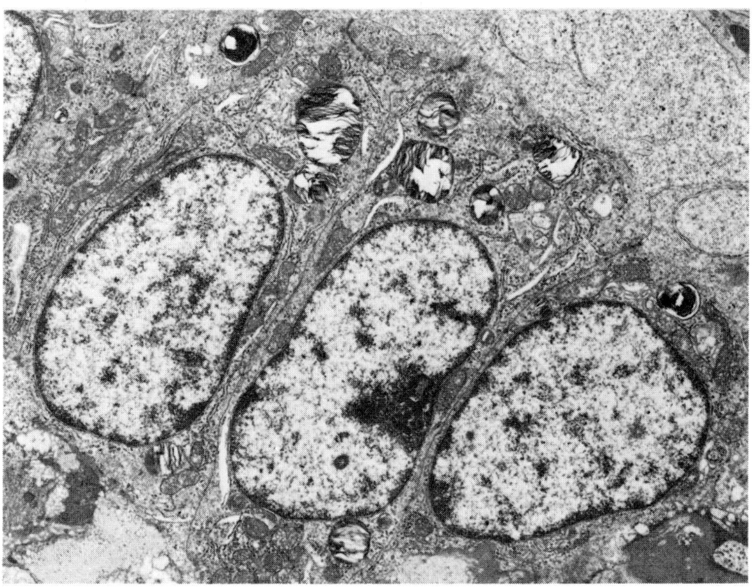

FIG. 2. Fine structure of the cancer cells showing features of type II pneumocytes. Double stain. X 5,000.

cells projecting into the bronchiolar lumen showed active apocrine secretion of the apical translucent cytoplasm.

Table 2

Hyperplastic and neoplastic change in lungs of rats exposed to nickel oxide

Experiment No	1 & 2			3			4		
Time sacrificed	Immediately after one month exposure			12 months after one month exposure			20 months after one month exposure		
Exposure level	High	Low	Control	High	Low	Control	High	Low	Control
Number of animals	7	5	7	5	5	4	8	6	5
Bronchiolar epithelialization	0	0	0	1	0	2	1	1	2
Bronchial gland hyperplasia	0	0	0	2	0	2	6	5	0
Squamous metaplasia	0	0	0	2	0	0	3	3	2
Adenomatosis	0	0	0	0	0	0	1	1	0
Adenocarcinoma	0	0	0	0	0	0	0	1	0

Hyperplastic and neoplastic lesions are listed in Table 2. Most of them occurred in aged rats of the 4th group. A case of pulmonary adenocarcinoma was detected in low-exposure-rats sacrificed at 20 months after one month inhalation of NiO. Electron microscopy revealed that the cancer cells showed mainly features of type II pneumocytes and partly of Clara cells (Figs 1 & 2).

Among cranial lesions of the 4th group, a kind of sarcoma in a high-exposure-rat was distributed from the paranasal area to the retroperitoneum. No carcinoma of the upper respiratory tract was found in any experimental group.

CONCLUSIONS

Carcinogenic effects of NiO aerosols were not prominent in these experiments, although the numbers of animals used were limited. Some neoplastic lesions in a few rats could possibly be associated with aging effects. Nevertheless, long residence of NiO particles in the peripheral lung might play a role in the promotion of bronchioloalveolar cell hyperplasia.

ACKNOWLEDGEMENTS

The authors wish to thank Miss Hiroko Noguchi, Miss Gyoko Zaitsu and Mr Mitsuru Yokoyama for their technical assistance.

REFERENCE

Tanaka, I. et al.(1982) Journal of the University of Occupational and Environmental Health, 4, 441-449.

ELECTRON MICROSCOPIC OBSERVATIONS ON LEIOMYOSARCOMAS INDUCED WITH NICKEL SUBSULFIDE

CARLO BRUNI

Department of Neurology,
University of Virginia Medical Center,
Charlottesville, Virginia 22908 USA

1. INTRODUCTION

The question of whether tumors induced by nickel compounds exhibit normal or abnormal cytodifferentiation has received only little attention. Riggs and Jones (1983), however, have reviewed the evidence that a number of chemical carcinogens (nickel has not been tested) interfere with one of the controlling mechanisms of normal cytodifferentiation. Such a result has thrown a new light on the hypothesis that abnormal differentiation by carcinogens might be a factor in the final emergence and in the propagation of some forms of tumors (Rubin, 1983). Consequently, it seems interesting to report that leiomyosarcomas (LMS) induced by the injection of nickel subsulfide (Ni_3S_2) into striated muscle were accompanied by abnormal differentiation.

2. MATERIAL AND METHODS

Ni_3S_2 (0.4 mg) in the form of dust was suspended in aqueous penicillin and injected into the muscle rectus abdominis of 6-week-old Fisher 344 male rats. Tissues for electron microscopy were fixed in 3% glutaraldehyde in phosphate buffer (pH 7.3), postfixed in 1% OsO_4, dehydrated in alcohols, and embedded in Araldite according to routine techniques.

3. RESULTS AND CONCLUSIONS

Seven of 31 animals that had received the carcinogen developed LMS at the site of injection. The latency period was 252-323 days.

In the electron microscopic image the tumors were composed of two classes of cells. A) Fibroblast-like cells that did not exhibit the filaments usually found in normal smooth muscle. B) Cells in which the appearance of filaments occurred in the following sequence: first, actin filaments (AF) in concomitance with intermediate filaments (IF); second, myosin filaments (MF). With time, AF and MF increased gradually in number, whereas IF increased rapidly at first and subsequently decreased.

Studies from other laboratories have reported that in the embryos of experimental animal species smooth muscle was derived from undifferentiated fibroblast-like cells and that with time they developed filaments in the same order as in the LMS cells. Tumor cells, in addition, shared similarities with normal cells in respect to the quantitative changes in filaments occurring during development.

However, dissimilarities were also detected. For instance, gap junctions, that are known to increase rapidly in number during normal smooth muscle development, were rare in LMS. Such a defect in maturation had a special significance in that gap junctions mediate the passage between adjacent cells of molecules that are involved in the control of cell growth in normal developing cell systems (Pitts, 1978). Another dissimilarity was that in the tumor cells, the smooth portion of the sarcoplasmic reticulum was far less developed than in the normal. In summary, it appeared that the differentiation of these tumor cells was initially similar to that of the normal prototypes, but that it became abnormal in the advanced stages.

CONCLUDING REMARK

The significance of the present observations is indicated by the consideration of two facts. The first is that induction of tumors by the intramuscular injection of Ni_3S_2 is the most commonly used experimental animal model for the study of the carcinogenicity of nickel. The second fact is that abnormal differentiation, as observed in the present Ni_3S_2-induced LMS, was also reported in a study of Ni_3S_2-induced rhabdomyosarcomas (Bruni and Rust, 1975). The recognition of the high complexity of the controls (DNA transcription, RNA processing, translation efficiency) of gene expression makes it unlikely that abnormal differentiation in tumors derived from two distinct cell types is a merely fortuitous event. Thus, studies directed to gain information about whether or not abnormal cytodifferentiation is causally involved with other types of lesions in the formation of tumors after intramuscular injection of nickel, might be essential to the further understanding of the mechanism of its carciogenic activity.

REFERENCES

Bruni, C. and Rust, J.N. (1975). Fine structure of dividing cells and of non-dividing, differentiating cells of nickel sulfide-induced rhabdomyosarcomas. Journal of the National Cancer Institute, 54, 687-696.

Pitts, J.D. (1978). Junctional communication and cellular growth control. In Feldman, J., et al. (Eds.), Intercellular Junctions and Synapses, pp. 61-79. Chapman and Hall, London.

Riggs, A.D. and Jones, P.A. (1983). 5-Methylcytosine, gene regulation and cancer. Advances in Cancer Research, 40, 1-30.

Rubin, H. (1983). Understanding cancer. Science, 219, 1170-1172.

TOXICOLOGY AND CARCINOGENESIS STUDIES OF NICKEL OXIDE, NICKEL SULFATE, AND NICKEL SUBSULFIDE: DESIGN OF STUDY AND CHARACTERIZATION OF NICKEL COMPOUNDS

JUNE K. DUNNICK*, C.W. JAMESON* and JANET M. BENSON**
*National Toxicology Program, National Institute of Environmental Health Sciences, Research Triangle Park, NC 27709 USA
** Lovelace Biomedical and Environmental Research Institute, Albuquerque, NM 87185 USA

INTRODUCTION

The National Toxicology Program has initiated toxicology and carcinogenesis studies of nickel oxide (NiO), nickel sulfate ($NiSO_4 \cdot 6H_2O$), and nickel subsulfide (Ni_3S_2) by inhalation exposure in Fischer 344/N rats and $B6C3F_1$ mice. Nickel oxide and nickel sulfate were chosen because of high exposure in the workplace (NIOSH, 1977, NAS, 1975) and nickel subsulfide, a known carcinogen in rodents after inhalation exposure as a positive control.

In 1976 the International Agency for Research on Cancer reviewed studies with nickel including animal data and epidemiology data in humans and concluded that nickel in some form is probably carcinogenic (IARC, 1976). It is still not possible to identify the specific nickel compounds which are carcinogenic in man (IARC, 1982). The NTP rodent studies are designed to help determine the relative human toxicities. nickel compounds to which man may be exposed.

CHEMICAL CHARACTERIZATION

Nickel oxide (CAS No. 1313-99-1) and nickel subsulfide (CAS No. 12035-72-2) were supplied by Inco, Ltd, Toronto, Ontario, Canada. and nickel sulfate (CAS No. 10101-97-0) was obtained from Aldrich Chemical Co. Nickel subsulfide was milled using a 4" micronizer (Sturtevant Mill Co., Boston, MA). The purity estimation of these compounds was based on elemental analyses and spark source mass spectroscopy (NTP, 1983 a,b,c; 1984 a,b,c).

The purities of the nickel oxide, sulfate and sulfide were found to be 99.5% and 97% respectively.

DESIGN OF GENERATING SYSTEMS

The generating systems were designed by the Lovelace Biomedical and Environment Research Institute, Albuquerque, NM (Carpenter et al., 1982). Mass median aerodynamic diameter (MMAP) was determined by cascade impactor (Hinds et al., 1982). A fluidized bed aerosol generator was used to produce respirable aerosols of nickel oxide and nickel subsulfide. A respirable aerosol of nickel sulfate was generated by nebulizing an aqueous solution of the compound using a pair of Retec nebulizers. In all cases, the mass median aerodynamic diameters of the aerosols generated were approximately 3 μM. In addition, the maximum aerosol concentrations proposed for the prechronic aerosol studies (50 mg/m^3 for nickel oxide and nickel sulfate and 4 mg/m^3 for nickel subsulfide) were attainable using these generation systems.

STUDY DESIGN

Fischer 344/N rats and $B6C3F_1$ mice will undergo whole body exposures 6 hrs/day, 5 days/week in Hazleton 1000 or 2000 inhalation chambers. These animal species are routinely used in the NTP toxicology studies, and a complete historical data base is available on the background tumor rates in these animals (Haseman et al.). Target sites after nickel exposure via the inhalation route are expected to include the lung and nasal cavity. The historical incidence of

TABLE 1. Summary of study design

Two-week repeated dose study
- -Male and Female Fischer 344/N Rats and $B6C3F_1$ Mice
- -Inhalation Exposure (50,25,12.5,6.25,3.1,1.5,0 mg/m^3)
- -Five Animals/Sex/Species/Dose Level

Ninety-day studies
- -Male and Female Fischer 344/N Rats and $B6C3F_1$ Mice
- -Inhalation Exposure (5 dose levels)
- -Ten Animals/Sex/Species/Dose Level

Chronic studies
- -Male and Female Fischer 344/N Rats and $B6C3F_1$ Mice
- -Inhalation Exposure (3 dose levels)
- -Fifty Animals/Sex/Species/Dose Level

tumors at these sites are low for Fischer 344/N rats and B6C3F$_1$ mice. The toxicology studies will consist of 14-day, 90-day, and two-year studies. Endpoints for evaluation include clinical signs, body weights, gross and microscopic examination of tissues, hemotologic measurements, and blood enzyme levels.

Special studies will be conducted to determine the correlation of toxicity of a particular nickel compound with the amount of nickel deposited in the animal and in a particular organ. There is evidence in both animals and humans that immunosuppression may enhance susceptibility to carcinogenesis, and nickel compounds have been shown to suppress the immune system. Studies will be conducted to determine the degree to which each nickel compound affects the immune system, and an attempt made to correlate any carcinogenic response with alterations in the immune system. A standard battery of immune tests which has been developed by the NTP will be used (Luster et al., 1982).

A prime concern in the development of methods to carry out these nickel toxicology studies is the development of good health and safety procedures. Each nickel study will be conducted in a separate room with air from each room vented through a series of HEPA filters. The nickel test material is prepared for use in a hood. Each generating system is contained within a hood and vented through two HEPA filters. Protective clothing and respirators are worn in all chamber rooms.

CONCLUDING REMARK

The nickel compounds to be investigated have different solubilities. Because of this we expect to find differences in the distribution of the compounds within exposed animals, and differences in the location and types of the resulting pathological effects. Results obtained from these studies may help us identify which nickel compounds contribute to the carcinogenicity observed in workers exposed to a variety of nickel-containing compounds.

REFERENCES

Carpenter, R.L. and Yerkes, K. (1980) Relationship between fluid bed aerosol generator operation and the aerosol produced. American Industrial Hygiene Association Journal 41, 888-894.

Haseman, J.K., Huff, J., and Boorman, G.A. Use of historical control data in carcinogenicity studies in rodents. Toxicology Pathology (in press).

Hinds, W.C. (1982) Aerosol Technology. Properties, behavior, and measurement of airborne particles. John Wiley & Sons, N.Y., N.Y. pp. 69-120.

International Agency for Research on Cancer. (1976) IARC Monographs on the Evaluation of the Carcinogenic Risk of Chemicals to Man. Cadmium, nickel, some epoxides, miscellaneous industrial chemicals and general considerations on volatile anaesthetics. 11,75-112.

International Agency for Research on Cancer. (1982) IARC Monographs on the Evaluation of the Carcinogenic Risk of Chemicals to Humans. Chemicals, industrial processes and industries associated with cancer in humans. IARC Monograph Supplement 4, 167-170.

Luster, M.I., Dean, J.H., Moore, J.A. (1982) Evaluation of immune function in toxicology. In Principles and Methods of Toxicology. ed. A.W. Hayes. Raven Press, N.Y. pp.561-586.

National Academy of Sciences. (1975) Medical and Biologic Effects of Environmental Pollutants. Nickel. Printing and Publishing Office. National Academy of Sciences. Washington, D.C. 20418.

National Institute of Occupational Safety and Health. (1977) Criteria for a recommended standard. Occupational exposure to inorganic nickel. DHEW (NIOSH) Publication No. 77-164, Washington, D.C.

National Toxicology Program Analytical Report. Chemical characterization of nickel sulfate hexahydrate. Midwest Research Institute, NIEHS Contract No. NO1-ES-95615, Sept. 16, 1983.

National Toxicology Program Analytical Report. Chemical characterization of nickel (II) oxide. Midwest Research Institute, NIEHS Contract No. NO1-ES-95615, Oct. 3, 1983.

National Toxicology Program Analytical Report. ASTM analysis of nickel oxide. Midwest Research Institute, NIEHS Contract No. NO1-ES-95615, Dec.5, 1983.

National Toxicology Program Analytical Report. Limited purity analysis of nickel subsulfide. Midwest Research Institute, NIEHS Contract No. NO1-ES-95615, Jan. 23, 1984.

National Toxicology Program Analytical Report. Spark source mass spectrometry of nickel subsulfide. Midwest Research Institute, NIEHS Contract No. NO1-ES-95615, March 14, 1984.

National Toxicology Program Analytical Report. Chemical Characterization of Nickel subsulfide. Midwest Research Institute, NIEHS Contract No. NO1-ES-95615. May 14, 1984.

461629

ALTERATIONS OF HUMAN FETAL KIDNEY EPITHELIAL CELLS PRODUCED BY NICKEL SULFATE

AAGE HAUGEN*, INGER-LISE HANSTEEN** and HELGE DALEN***
*Department of Toxicology, National Institute of Public Health, 0462 Oslo 4, Norway
**Department of Occupational Medicine, Telemark Sentralsjukehus, Porsgrunn, Norway
***Laboratory of Clinical Electron Microscopy, University of Bergen, Bergen, Norway.

INTRODUCTION

The occupational and environmental hazards of metal exposure are of great concern. Epidemiological evidence has implicated nickel compounds as the causative agents in respiratory and renal cancers in nickel refinery workers (Doll et al., 1977; Pedersen et al., 1973; Lessard et al., 1978). Nickel subsulfied induces renal carcinomas following intrarenal injection in rats (Sunderman et al., 1979) and nickel carbonate has been shown to induce protein-DNA cross-links in rat kidney in vivo (Ciccarelli et al., 1981). Furthermore, some metals such as cadmium, mercury, lead, chromium and nickel may have profound effects on fetal development.

The nickel ion is thought to be the ultimate carcinogenic form of nickel. Metal ions such as Ni^{2+}, Cr^{2+} and As^{5+} have been shown to be efficient transforming agents for animal cells in culture. In the present investigation we have studied the effects of nickel ions on human fetal kidney explant cultures. Our results demonstrate that Ni^{2+} will induce phenotypically altered human fetal kidney epithelial cells expressing altered growth requirements and karyology, extended population-doubling potential and growth in soft agar.

EPITHELIAL OUTGROWTHS FROM EXPLANT CULTURES

Human fetal kidneys from second trimester spontaneous abortions were cut into 0.3 x 0.3-cm pieces and transferred to tissue culture dishes containing LHC-medium. The explants were exposed continually to a semitoxic dose of nickel sulfate (5 μg/ml). The kidney epithelial cells grow off the exposed and control explants and established primary cultures. The explants were replanted after 14 days to produce a second outgrowth. This replanting prosess was repeated up to 10 times. In primary cultures derived from 16 of the 20 Ni-exposed explants, colonies of mitotic and morphologically altered epithelial cells appeared approximately after 70 to 100 days. These cells could be repeatedly subcultured. The control cultures became quiescent after five (1:3) subculturings in LHC-medium.

EFFECTS OF NICKEL ION ON GROWTH OF NORMAL FETAL KIDNEY EPITHELIAL CELLS

A partial characterization of colonies that arose in Ni^{2+}-treated cultures has been initiated. Treatment with Ni resulted in cell populations with an enhanced growth potential as compared to control cultures. Some of the cloned isolates have been subcultured 15 times with a minimum of 8-10 population doublings per passage. Thus compared with unexposed control cells, cultures isolated after prolonged exposure to Ni exhibit an increase in their population-doubling potential of at least 100 cell divisions.

Anchorage-independent growth was also examined. After 7 months primary cultures established from exposed explants were able to form colonies in soft agar. The colony-forming efficiency of carcinogen treated cells in soft agar ranged from 0.02 to 0.05% for different cell lines. Solvent treated controls were not able to form colonies in soft agar.

Typical examples of the morphology of some of the primary cell cultures are shown in figure 1.

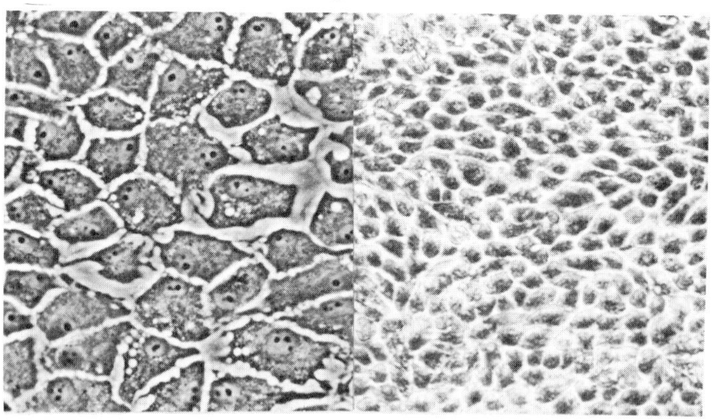

Fig. 1. Phase-contrast view of (a) crowded normal and (b) phenotypically altered cells (x 200).

In contrast with cultured kidney epithelium from normal explants, epithelial cells derived from carcinogen-exposed tissue had a markedly increase in nucleus:cytoplasmic ratio and the cells remained tightly packed. Electron microscopic observations confirmed the epithelial character of the cell-lines demonstrating such features as microvilli, tonofilaments, cilia and normal junctional complexes (fig. 2).

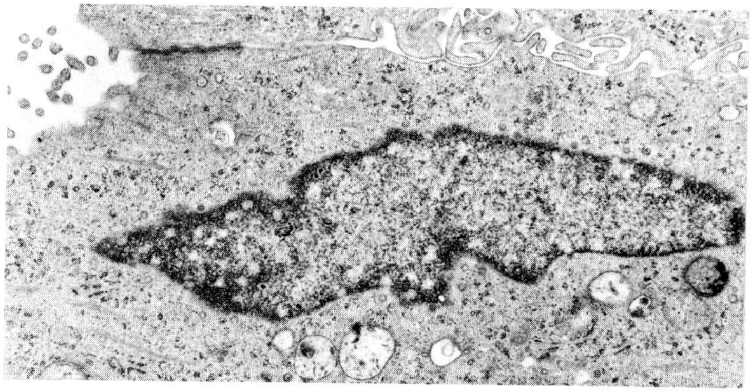

FIG. 2. Electron micrograph of phenotypically altered cells. (x 12.000).

Epidermal growth factor (EGF) is a mitogenic polypeptid in several in vitro systems. As seen in figure 3 EGF stimulates normal kidney epithelial cell proliferation. The effect of EGF on the growth rate of the phenotypically altered clones was also quantitated. Normal kidney epithelial cells were more responsive to EGF than were the isolates.

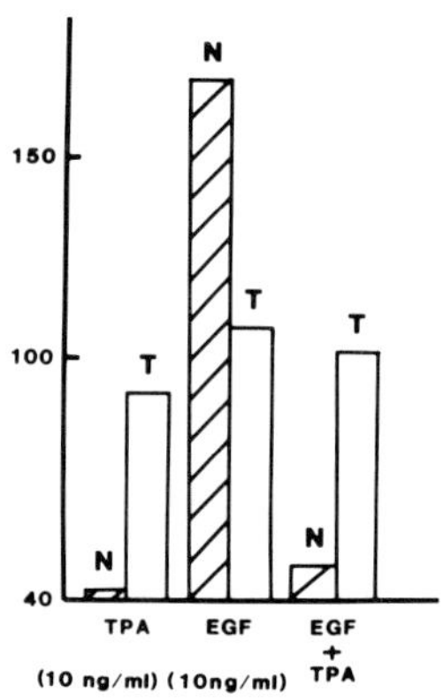

FIG. 3. Response of normal (N) and phenotypically altered (T) cells to TPA, EGF, or both.

In addition a differential effect of 12-0-tetradecanoyl-phorbol-13-acetate (TPA) was also noted. TPA is a potent tumor promoter which can either stimulate or inhibit the proliferation of cultured cells (Scribner and Suss, 1978; Diamond et al., 1974; Kinzel et al., 1981). As seen in figure 3 the isolates continued to proliferate in TPA-containing medium, while 10 ng/ml TPA inhibited growth of normal cells by 40-45%. Because of the relationship between EGF and TPA the growth-promoting activities of EGF in combination with the tumor promoter were also studied. TPA inhibited growth by 115-120% in EGF treated normal kidney epithelial cells and 5-10% in phenotypically altered cells. The inactive analog 4α phorbol had no significant effect. Karyotype changes have also been noted. Many isolates had a substantial subtetraploid component of cells, though this was not a necessary condition for establishment of indefinite lifespan since one phenotypically altered culture was found to be diploid when isolated. Marker chromosomes were also observed.

CONCLUSIONS

A culture system has been used to explore the response of human fetal kidney epithelial cells to Ni^{2+}. The phenotypically altered cells were characterized by reduced growth factor requirement for cell division, resistance to the inhibitory effect of TPA, growth in soft agar and chromosomal aberrations.

ACKNOWLEDGEMENT

Supported by NTNF Grant AM-10.10923.

REFERENCES

Ciccarelli, R.B., Hampton, T.H. and Jennette, K.W. (1981) Cancer Lett., 12, 349-354.

Diamond, L., O'Brien, S., Donaldson, C. and Shimazu, Y. (1974) Int. J. Cancer, 13, 721-730.

Doll, R., Mathews, J.D. and Morgan, L.G. (1977) Br. J. Ind. Med., 34, 102-105.

Kinzel, V., Richards, J. and Stohr, M. (1981) Cancer Res., 41, 300-305.

Lessard, R., Reed, D., Maheux, B. and Lambert, J. (1978) J. Occup. Med., 20, 815-817.

Pedersen, E., Hogetveit, A.C. and Andersen, A. (1973) Int. J. Cancer, 12, 32-41.

Scribner, J.D. and Suss, R. (1978) in Richter, G.W. and Epsteins, M.A. (Eds.), International Review of Experimental Pathology, Vol 18, pp. 639-648. Academic Press, New York.

Sunderman, F.W., Jr. (1978) Fed. Proc. Fed. Am. Soc. Exp. Biol., 37, 40-46.

Sunderman, F.W., Jr., Maenza, R.M., Hopfer, S.M., Mitchell, J.M., Allpass, P.R. and Damjanov, I. (1979) J. Environ. Pathol. Toxicol., 2, 1511-1527.

461630

BIOLOGICAL CHARACTERIZATION OF CELLS CULTURED FROM A SINONASAL CARCINOMA OF A FORMER NICKEL WORKER.

ALBRECHT REITH, RAGNAR VOSS, JORUNN JACOBSEN and MORTEN BOYSEN

Departments of Pathology and Biochemistry, The Norwegian Radium Hospital and Norsk Hydro's Institute for Cancer Research, and Ear, Nose and Throat Department, Rikshospitalet, Oslo, Norway and the Norwegian Cancer Society.

INTRODUCTION

So far, toxicological and biochemical investigations of nickel-induced cancer cells have been restricted to tissue material derived from laboratory animals (Reith and Brøgger, 1984). Recently developed methods of culturing monolayers and multicellular spheroids of human tumours for clinical screening of chemotherapeutric drugs make large quantities of **human** tumour cells available for research purposes.

We report here on the first successful cultivation of tumour cells from a poorly differentiated sinonasal carcinoma in a nickel worker. The cells were grown for several weeks on agar and gave rise to spheroids up to 300 µm in diameter. They produced a mucous material. Electron microscopy (EM) showed mainly two types of cells: Large cells with abundant organelles in the cytoplasm, especially greatly-extended rough endoplasmic reticulum (RER), and small cells with sparse organelles and abundant free ribosomes. Their surfaces were covered by pleomorphic microvilli and/or foldings.

MATERIAL AND METHODS

The material was derived from a tumour in the right nasal cavity of a 72-year-old retired worker who had been occupationally exposed to nickel in the roasting and smelting department of a refinery from 1949 to 1956. The tumour was diagnosed as a small cell tumour of epithelial origin. Before treatment, small tumour pieces were removed and either kept in a culture medium at 4°C or fixed in 3% cacodylate-buffered glutaraldehyde for EM.

For **cell culture**, the method of Wibe et al. (1984) was used. Tumour tissue was minced with crossed scalpels in Ham's F12 medium supplemented with 15% newborn calf serum. Cells and small pieces of tissue were cultured for 2-3 days before being transferred to agar-coated flasks for incubation at an atmosphere of 5% CO_2 in air. The first spheroids were observed after 3-4 days. After trypsinization the cells were subcultured up to 15 passages before being frozen and kept in liquid nitrogen for further analysis. Drug sensitivity to bleomycin and adriamycin was demonstrated.

The material for EM was dehydrated in graded ethanol, critical point dried for scanning EM (SEM) or Epon-embedded for transmission EM (TEM).

RESULTS AND DISCUSSION

EM demonstrated 2 types of cells in the biopsies: Cells with (i) sparse cytoplasm containing few mitochondria and abundant free ribosomes (Fig. 1a,b), and (ii) Golgi profiles and mucus granules and a well-developed RER filled with an amorphous material of appreciable density (Fig. 1c,d). Culturing of the bioptic material preserved these 2 main cell types, as seen in TEMs of the spheroids which show typical tumorigenic small cells (Figs. 2a and 3) and larger, secretorily active cells (Fig. 2b) with extended RER filled with electron dense material (inset), and a surface with pleomorphic microvilli and/or foldings (Fig. 4). Sometimes, flat, squamous-like cells were seen on the filter used in collecting the spheroids.

The ultrastructure resembled the carcinoma types in nickel workers with respect to cytoplasmic fine structure and SEM surface characteristics described recently by us (Reith et al., 1983). Adenocarcinoma was characterized by abundant RER, Golgi profiles and microvilli-covered intercellular spaces, whereas poorly differentiated carcinoma displayed sparse organelles but abundant free ribosomes.

Culturing of the mechanically-treated bioptic material resulted in monolayers of cells partly outgrowing from small tissue pieces. Culturing on agar resulted in loosely structured spheroids which could be subcultured up to 15 passages. The doubling time was about 24 hours. A mucus production was simultaneously observed, characterized by the culture medium becoming slimy.

The multicellular spheroid is the in vitro culture system which is biologically closest to the primary tumour and reflects its properties best, as shown for a variety of tumours (Wibe et al., 1984). EM demonstrates very clearly the heterogeneity of the primary tumour material. This heterogeneity was still maintained after culturing up to 15 passages and makes this cell system more suited for biochemical and toxicological studies than the mostly used permanent cell lines, which no longer necessarily reflect the characteristics of the primary tumour cells.

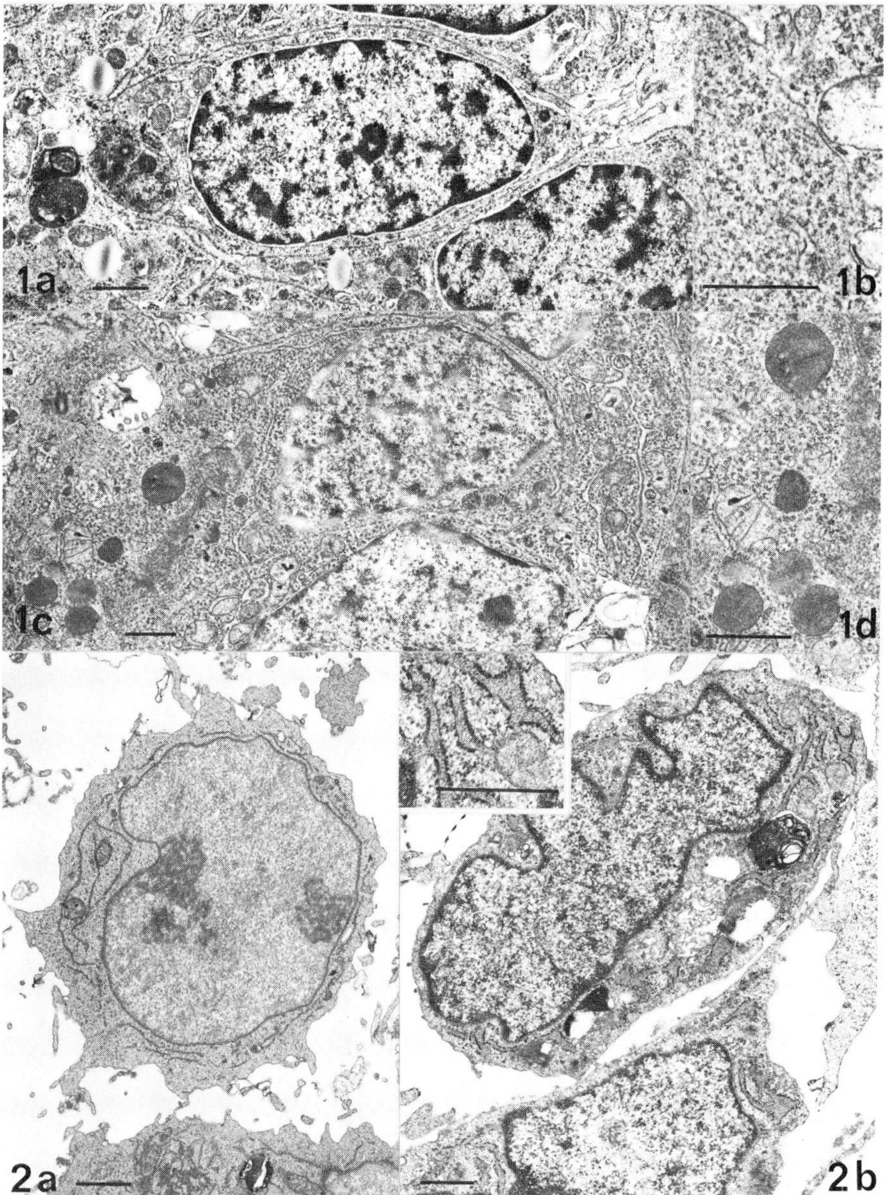

TEM of sinonasal carcinoma (bar=1 µm). Fig. 1: Primary tumour with small cells (a) with abundant free ribosomes, (b) and secretory cells (c) with mucus granules and RER (d). Fig. 2: Cultured spheroids with small cells (a) and secretory cells (b) with extended RER (c).

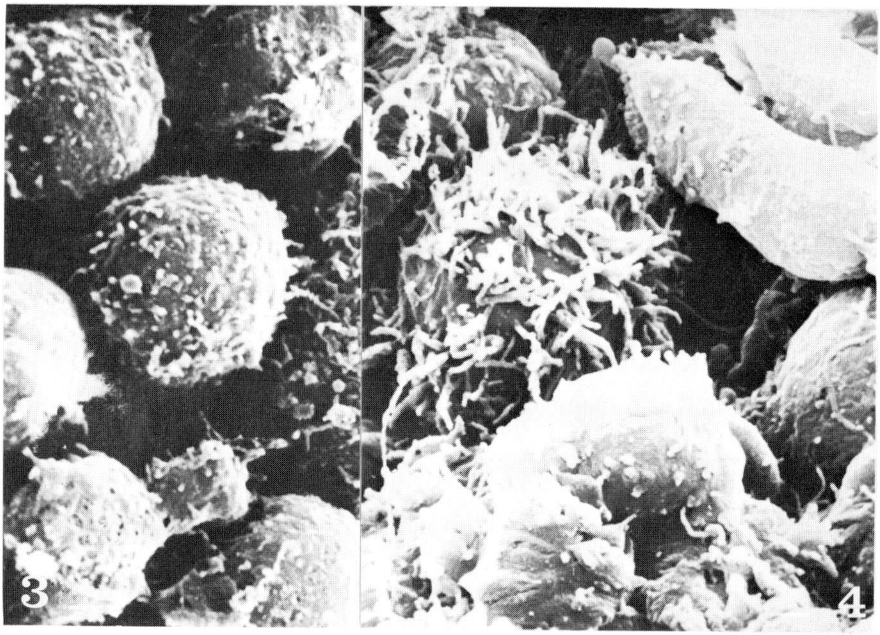

SEM of cultured sinonasal carcinoma cells. Small cell tumour cells with sparse, knob-like microvilli (Fig. 3) and larger, probably secretory cells with pleomorphic microvilli and foldings (Fig. 4). (bar = 1 μm).

CONCLUDING REMARK

Karyotyping will be performed on these cells; they are offered to other investigators who are interested in nickel carcinogenesis.

REFERENCES

Reith, A. and Brøgger A. (1984) in Nickel in the Human Environment. Proceedings of the symposium on Nickel in the Human Environment. Lyon, 8-11 March 1983. IARC in press.

Wibe, E., Berg, J.P. Tveit, K.M., Nesland, J.M. and Lunde, S. (1984). International Journal of Cancer 34, 21-26.

Reith A, Rigaut J.-P., Boysen, M. and Marton, P. (1983). in Reznick G. and Stinson S.F. (Eds.). Nasal Tumors in Animals and Man, vol. II, pp. 55-94. CRC Press Inc., Boca Raton, Florida, USA.

461631

INCORPORATION OF αNi_3S_2 AND βNiS INTO HUMAN EMBRYONIC PULMONARY CELLS IN CULTURE.

HARTMUT F.HILDEBRAND*, MAUD COLLYN D'HOOGHE** and MARIE-CLAIRE HERLANT-PEERS**
* Institut de Médecine du Travail, 1 Place de Verdun. 59045 Lille-Cédex,France.
** INSERM U.124, 1 Place de Verdun, 59045 Lille-Cédex,France.

INTRODUCTION

Clinical pathological and experimental investigations have shown that lung is the target organ of Ni-incorporation (Sunderman and Sunderman, 1963; Herlant-Peers et al. 1983) and that Ni-compounds mostly induce pulmonary carcinoma (Pedersen et al. 1973). One of the most carcinogenic Ni-compound is αNi_3S_2, whereas the carcinogenicity of βNiS is estimated to be very low. The toxic effect and the incorporation of both compounds are studied on human embryonic epithelial pulmonary cells in culture (L132).

MATERIAL AND METHODS

L132 cells derived from normal human embryonic lung epithelium are cultured in plastic bottles in MEM 0011 medium (Eurobio) supplemented with 5% fetal calf serum.

In order to establish growth curves after continuous exposure to βNiS and αNi_3S_2, $1.5x10^5$ growing cells were seeded in 2ml Leighton tubes and incubated at 37°C for 8 days. Every day the cells were suspended in 0.1% trypsin and the number of cells per tube was measured. Four tubes were used for each point and at least three seperate experiments were realised for each concentration of the substance tested. The survival rate was determined using the colony forming method.

Ultrastructural investigation was carried out on cells grown in the presence of 50µMβNiS and αNi_3S_2 for 3 days. The material was prepared for electron microscopy as described previously (Hildebrand and Biserte, 1979). EDS microprobe analyses were performed on unstained sections placed on

Pioloform F coated titanium grids. Elemental determinations were carried out with a Philips EM 301 electron microscope and a Link-System microprobe analyser.

RESULTS AND DISCUSSION

Cellular growth. The growth curves were established for different concentrations of βNiS (Fig.1) and αNi_3S_2 (Fig.2). At equal molarity, the βNiS had a much higher inhibitory effect on the cell growth than the αNi_3S_2. The same deceleration effect for both is 200 µM and 400 µM respectively.

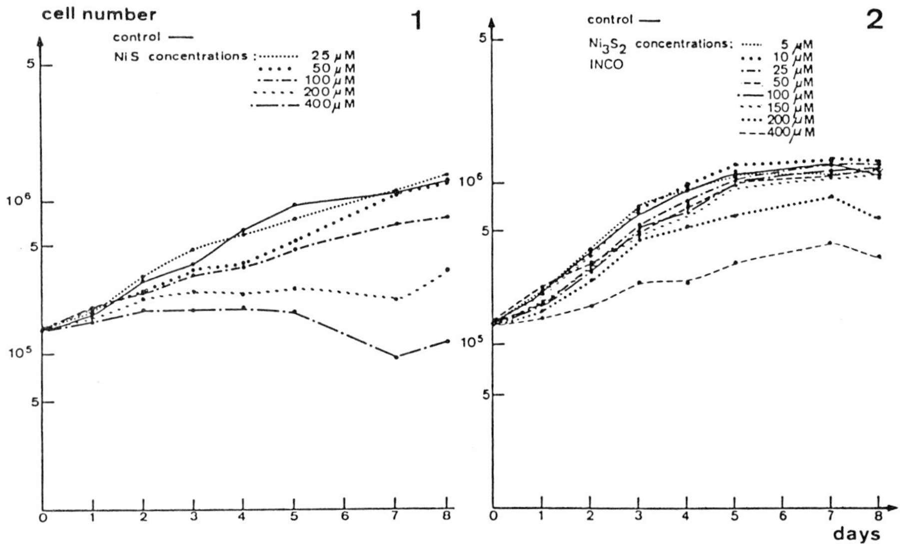

Growth curves of L132 cells for different concentrations of βNiS (Fig.1) and αNi_3S_2 (Fig.2)

The higher cytotoxic effect of βNiS is confirmed after establishing the survival curve by the colony forming method (Fig.3). Both compounds were compared with $NiCl_2$ and $NiCO_3$ which were found to be less toxic, although the inhibitory effect of $NiCO_3$ on cell growth is much higher than that one of $NiCl_2$ (Collyn d'Hooghe et al.1983). These results show clearly that the cytotoxic effect of Ni-compounds is not directly related to their carcinogenic or mutagenic effects.

The ultrastructural study of the cells cultured in presence of the compounds at identical concentrations reveals important changes in the shape and

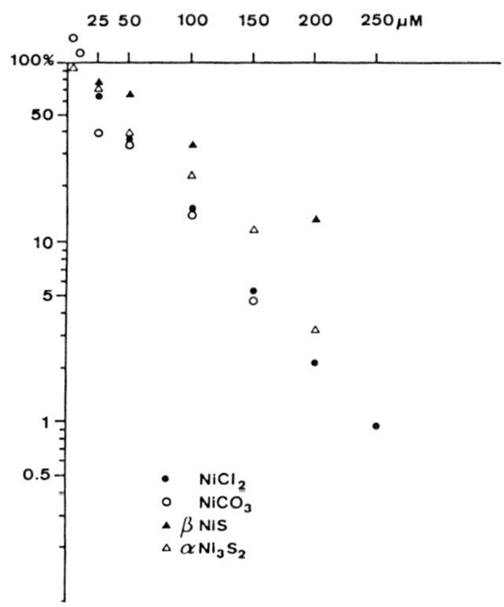

Fig.3 Survival curves of L132 cells cloned in presence of different Ni-compounds.

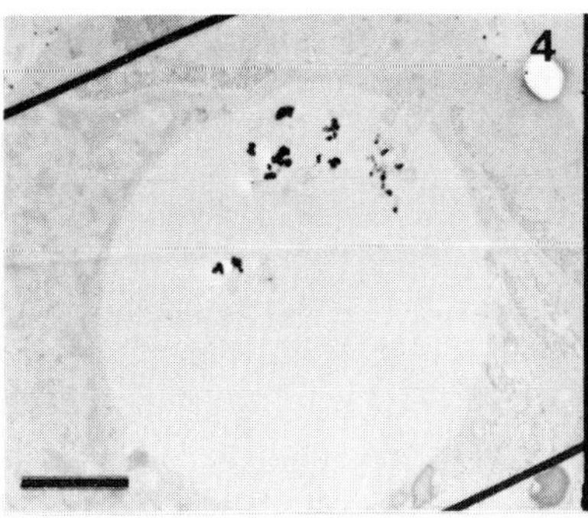

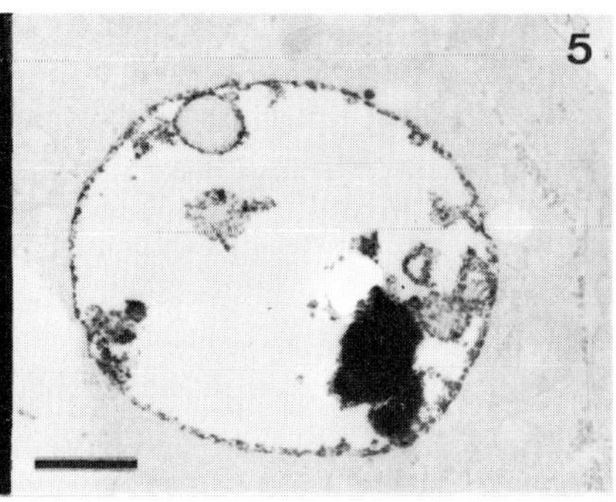

Fig.4 βNiS particles in an intracellular vacuole of L132 cells after a 3-day culture.bar:1µm.
Fig.5 Fine membrane bound αNi_3S_2 particles in an intracellular vacuole of L132 cells after a 3-day culture.bar:0.5µm.

the size of the cells as well as of their nuclei. Whereas untreated cells are large, smooth shaped with clear and regular nuclei, Ni-treated cells are retracted, globular and with a high number of microvilosities. They often contain large strands of glycogen granules. Their nuclei are irregular shaped and chromatin strands are more condensed than in untreated cells.

The incorporation of the two compounds occurs in a very different way. βNiS is generally found under its original fragmental form in large intracellular vesicles (Fig.4). αNi_3S_2, however, is recovered as small particles with an average size of 10 µm generally bound to the outer cell membrane, to the membrane of intracellular vesicles and to lysosomal structures (Fig.5). In a previous study we already showed the membrane fixation of $NiCO_3$-complexes (Collyn d'Hooghe et al. 1983).

The EDS microprobe analysis of the membrane- and the lysosome bound particles reveals besides Ni and S the presence of P. This observation suggests the formation of a Ni/P complex on the membranes.

CONCLUSIONS

Three main observations of this investigation can be pointed out :

The cytotoxic effect of nickel compounds is not directly related to their carcinogenic or mutagenic action.

Carcinogenic compounds such as $NiCO_3$ and Ni_3S_2 have a special affinity to cytoplasmic and cell membranes of epithelial pulmonary cells in culture.

The present investigation and our previous observations on $NiCO_3$ incorporation strengthen the metabolization hypothesis of Ni-compounds.

REFERENCES

Collyn d'Hooghe M., Kerckaert J.P., Herlant-Peers M.C.and Hildebrand H.F.(1983) in Anke M.,Baumann W., Bräünlich H.and Brückner C. (Eds) 4.Spurenelement Symposium pp 45-52 Jena.

Herlant-Peers M.C., Hildebrand H.F. and Kerckaert J.P. (1983). Carcinogenesis 4, 387-392.

Hildebrand H.F. and Biserte G. (1979) European Journal of Cell Biology 19, 276-280.

Pedersen E., Høgetveit A.C. and Anderson A. (1973) International Journal of Cancer 12, 32-41.

Sunderman F.W.Jr and Sunderman F.W. (1963) American Journal of Clinical Pathology, 40, 563-575.

ACKNOWLEDGEMENTS

Grants from the Institut National de la Santé et de la Recherche Médicale (INSERM U.124) and the Association PROTHESOR are acknowledged.

SUMMARY REMARKS OF SESSION CHAIRMEN (F. William Sunderman Jr., Gaetan Jasmin, Albrecht Reith, and Hartmut Hildebrand)

Dr. Gilman's Plenary Lecture reviewed two decades of experimental studies on nickel refinery dust, and its components, which support the epidemiological evidence that there is a causal relationship between occupational exposures and upper respiratory tract and lung cancers. The induction of muscle tumors by nickel has provided a useful model in studies of numerous aspects of tumorigenesis, including species susceptibility, strain differences, and the influence of a compound's solubility on tumorigenic response. That there is still much to learn was demonstrated by the papers presented by Drs. Bruni, Jasmin and Lumb, which dealt with the types of tumors induced by nickel compounds and the origins of the tumoral stem cells.

The later papers in these sessions concentrated on the carcinogenic potency of different nickel compounds and the modulation of the underlying process. By contrast to former studies, in which nickel compounds were just tested for tumour outcome, the models presented in these papers were essentially oriented to a better understanding of the biology of the carcinogenic process. A trend was evident in the use of models corresponding more closely to real exposure situations and target organs, including the applications of human cell lines.

The modulation experiments of Dr. Kasprzak indicate that magnesium, but not calcium, inhibits nickel carcinogenesis. This inhibition only occurs after local concurrent i.m. administration of magnesium and nickel compounds, and it occurs in spite of a very short retention of magnesium, in contrast to nickel, which is retained much longer. The results of this *in vivo* test correlate well with formerly reported tests of the half-life of nickel compounds *in vitro*, and they complement previous investigations by F. W. Sunderman Jr. and his co-workers on the inhibitory effect of manganese upon nickel carcinogenesis. The key question now is whether these findings can be extended to other organs.

This good correspondence between *in vivo* and *in vitro* studies was also shown for the transformation of kidney cells by nickel compounds. Making use of human fetal kidney epithelial cells in culture, Dr. Haugen demonstrated that with carefully chosen conditions, nickel sulfate induced malignant transformation with chromosomal aberration.

A new model of tumour induction was presented by Dr. Skaug who performed intrapleural administration of nickel subsulphide and nickel oxide and made comparisons with asbestos.

Although the same administration procedure was used for all tested compounds, the tumoral stem cells for nickel-induced tumours and asbestos-induced tumours seem not to be the same. It was suggested that there is a different mechanism of tumour induction, as reflected by histological differences between the tumours which were obtained. Dr. Horie presented some interesting results on inhalation studies with nickel oxide. The principal lesions observed at an interval of one year after a period of one month's exposure showed hyperplasia without dysplasia. Nickel oxide was incorporated into macrophages and in mucous cells, and it was still present in pockets of terminal bronchioli. No nickel particles were observed in pulmonary epithelial cells.

In order to assess more precisely nickel carcinogenesis in the human situation, the strategy and preliminary findings of a detailed animal research programme were presented by Dr. Dunnick. This paper engendered a lively discussion, which focussed on the need for considering the particle size, for insoluble compounds, and also the carcinogenic potential of compounds such as nickel carbonyl.

TOXICOLOGY

X-RAY MICROANALYTICAL AND AUTORADIOGRAPHIC DETECTION OF NICKEL IN BACTERIAL CELLS

DAVID C. SIGEE, RAZZAK N. AL-RABAEE and HISHAM A. EL-MASRY

Cytology Unit, Departments of Botany and Zoology,
University of Manchester,
Manchester M13 9PL UK

INTRODUCTION

X-ray microanalysis and autoradiography provide useful experimental tools for investigating the uptake and occurrence of nickel within intact cells. In this paper, the use of these techniques is described for the detection and localisation of nickel in cells of the bacterium *Pseudomonas tabaci*, grown at low (not exceeding 1ppm) external levels of the cation.

X-RAY MICROANALYSIS

Specimen preparation

Bacteria were grown in nutrient broth (containing less than 10 μM Ni), washed in buffer, and either placed directly on an EM formvar-grid (unfixed cells) or fixed in 2.5% glutaraldehyde followed by ethanol dehydration (fixed cells). X-ray microanalysis, using a 0.3μm diameter probe spot, was carried out on whole cells using similar equipment and operating conditions to those described previously (Sigee & Kearns, 1980).

Results and conclusions

Figure 1 shows the appearance of bacterial cells on the EM grid. Representative X-ray emission spectra from unfixed and fixed cells, with formvar control, are shown in Fig. 2. In the fresh (unfixed) sample the soluble ions K and Cl are very prominent - but transition metals Ni, Cu and Zn cannot be detected. Chemical fixation and dehydration results in the loss of soluble constituents, with the retention of insoluble Ni, Cu and Zn. The ability to detect nickel and other

Fig. 1 Transmission electron microscope view of whole bacterial cells (glutaraldehyde fixed, dehydrated on formvar). Bar equals 1 micrometre.

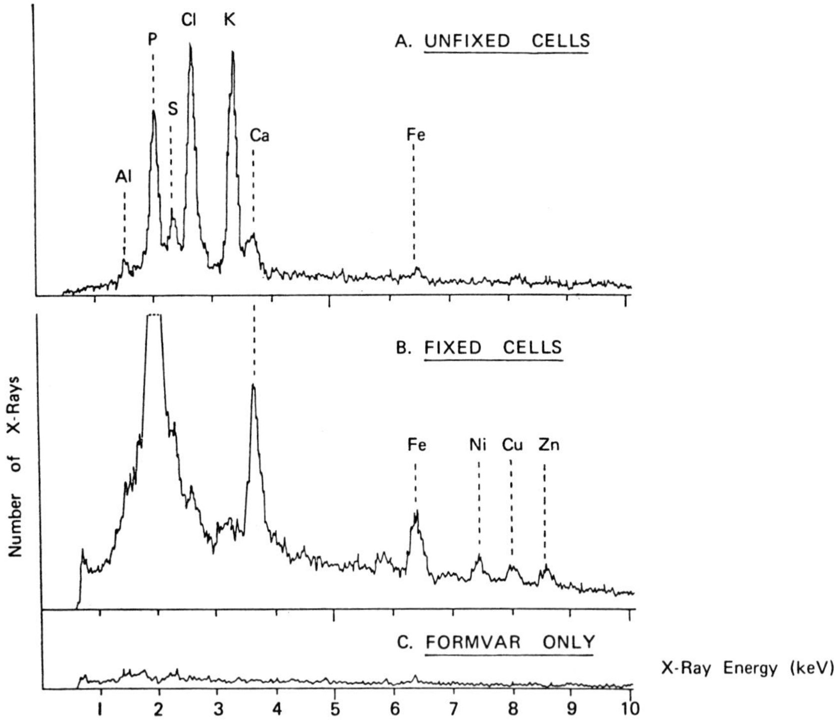

Fig. 2 X-ray Emission spectra - taken from (a) whole unfixed (fresh) cells, (b) whole glutaraldehyde-fixed, dehydrated cells, and (c) region of clear formvar (control) adjacent to bacteria.

transition metals more readily in the fixed preparation is attributed to two main factors - a) a greater peak signal in the probe area due to increased nickel concentration arising from cell shrinkage during dehydration, and b) a relative increase in the nickel mass fraction during processing due to selective loss of much of the cell matrix (particularly soluble proteins). The clear nickel peak seen after fixation was present in logarithmically-growing cells (Figs. 1 & 2) but not stationary phase cells. Mass fractions were highly variable, with a mean value at approximately 10 µmole Ni/g of material.

AUTORADIOGRAPHY

Specimen preparation

Bacteria were grown for 2h in $^{63}Ni^{2+}$ (8.3µCi/ml, total concentration 20 µM Ni) then processed for scintillation counting and autoradiography as detailed by Al-Rabaee and Sigee (1984).

Results and conclusions

Scintillation studies showed that only 15% of the radioactive nickel taken up by the bacteria occurred as an insoluble constituent. The location of this bound nickel was determined by autoradiography.

Figure 3 shows a high resolution EM autoradiograph, which has been processed by gold latensification and physical development. The very fine silver grains (mean diameter 0.04 µm) show clear localisation to regions of bacterial chromatin precipitated in the centre of the cell, with no apparent retention of insoluble nickel in peripheral cytoplasm or cell wall.

GENERAL DISCUSSION

The two microscopical techniques used here to detect nickel in bacteria differ in the information that they can reveal. Autoradiography detects *nickel uptake* (dynamic occurrence), while X-ray microanalysis detects *total nickel level* (static occurrence). The two techniques also differ in their detection sensitivity, with autoradiography able to detect an estimated 10^{-20}g Ni within a 0.2 µm² area of section compared to about 10^{-16}g Ni for X-ray microanalysis (Sigee and Kearns, 1980).

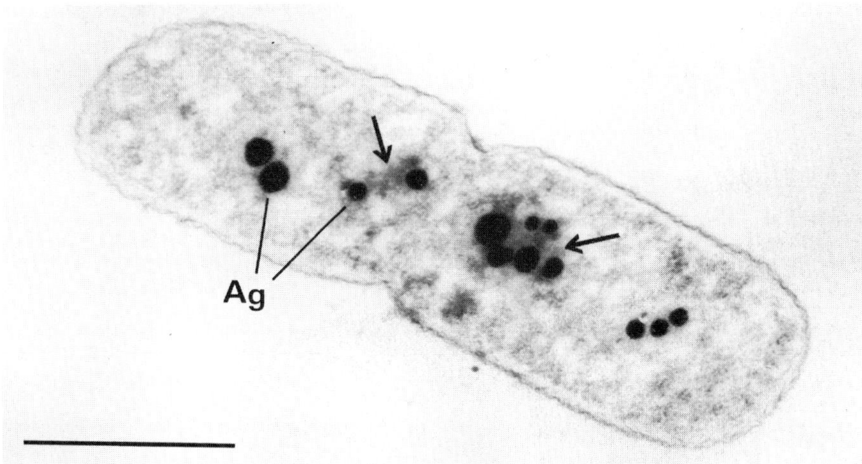

Fig. 3. EM autoradiograph. Section of acetic-alcohol fixed cell. Ag- silver grains; arrows - chromatin. Bar equals 0.5 micrometres.

CONCLUDING REMARK

The localised occurrence of insoluble nickel to bacterial chromatin is of considerable interest since it parallels the situation seen in dinoflagellate cells (Sigee, 1982). This provides further support for attributing a prokaryote affinity to this group.

REFERENCES

Al-Rabaee R.N. and Sigee D.C. (1984). Journal of Cell Science (in press)

Sigee D.C. (1982). Protoplasma 110, 112-120.

Sigee D.C. and Kearns L.P. (1980) in Brown S.S. and Sunderman F.W. (Eds), Nickel Toxicology, pp. 163-166. Academic Press, London.

461632

SOLUBILITY OF METALLIC NICKEL AND NICKEL OXIDES IN A BIOLOGICAL FLUID AT DIFFERENT pH LEVELS

KAREN HANSEN

Danish National Institute of Occupational Health,
Baunegaardsvej 73, DK-2900 Hellerup, Denmark

INTRODUCTION

The carcinogenicity of black and green nickel oxides and of metallic nickel dust has not been fully clarified. Six different compounds were obtained from INCO Ltd.: black and green NiO, two metallic nickel dusts and two ferronickel alloys. The compounds had been tested in the BHK-21 cell transformation assay, and the results showed that at the 50% toxic level all compounds had essentially the same transforming effect. The active concentrations, however, were within a wide range (unpubl. results). It has previously been shown that the transforming potency of soluble as well as of insoluble nickel compounds depends on the amount of available nickel (Hansen and Stern 1983).

The solubility of the nickel compounds is one of the elements determining the intracellular availability. Small particles (< 5 μm) of water-insoluble nickel and nickel oxides are phagocytized by mammalian cells in culture (Kuehn et al. 1982; Costa et al. 1981). After phagocytosis the particles are solubilized inside the cells in vacuoles (Abbracchio et al. 1982) in which the pH level has been shown to become acidic upon ingestion of particulates (Jensen and Bainton 1973). As the solubility of nickel compounds depends on the medium of dissolution as well as on the pH level the amount of available nickel could be related to the solubility in an appropriate medium of an acidic pH level. Therefore the solubilization of seven nickel compounds in serum was determined at three different pH levels: 4.5, 6.0 and 7.5; and the results were related to the relative toxicities of the the compounds determined using the BHK-21 in vitro toxicity assay (Hansen and Stern 1983).

MATERIALS AND METHODS

Nickel compounds.

The nickel compounds tested were: Sample 1) Black nickel oxide NiO purchased from BDH Chem. Ltd; Sample 2) Black nickel oxide NiO, produced by thermally decomposing basic nickel carbonate; Sample 3) Nickel powder produced by decomposing $Ni(CO)_4$; Sample 4) Nickel powder produced by reducing nickel ammonium sulphate in solution with hydrogen under pressure; Sample 5) Ferronickel alloy powder produced by decomposing a mixture of nickel and iron carbonyls; Sample 6) Ferronickel alloy (coarse grains); Sample 7) Green nickel oxide NiO produced by oxidizing Ni_3S_2 at high temperatures. Sample 2)-7) were obtained from INCO Ltd. together with the information about the compounds.

Dissolution tests.

The procedure was essentially as described by Kuehn and Sunderman Jr (1982). For each of the nickel compounds an amount corresponding to 8 mg of nickel was suspended in 4 ml of fetal calf serum in screw-cap tubes (2 for each compound). The serum had been buffered (phosphate buffered saline) and adjusted to pH 4.5, 6.0 and 7.5. After 1h, 3h, 6h, 12h, 24h, 48h and 72h of incubation at 37°C in an oscilliating waterbath the tubes were centrifuged and 50 µl were removed from each tube for analysis. After adding 3 ml of pure water to the 50 µl, the samples were centrifuged using silicone gel caps (Sure-Sep II) to ensure that no particles of undissolved material were retained in the samples.

Atomic absorption spectroscopy (AAS).

The samples were analyzed by flame-AAS with 0.5 ppm as the detection limit. Those samples which had a lower nickel content were analysed by electrothermal AAS (Perkin-Elmer 2380) using a graphite furnace atomizer (detection limit 3-5 ppb). No contamination above this level was detected in any of the corresponding negative controls. The injection volume was 20 µl and the temperature programme was as follows: (a) increase from 20°-120°C in 5 sec, (b) drying of sample for 30 sec at 120°C,(c) increase to 1000°C in 10 sec, (d) ashing of sample for 5 sec at 1000°C, (e) atomization for 5 sec at 2500°C and (f) 3 sec at 2650°C to clean the graphite tube. To enhance the sensitivity the internal flow of argon was reduced from 300 ml to 50 ml/min during the atomization step.

Toxicity tests.

Baby hamster kidney cells (BHK-21 cell line) were seeded in 60 mm dishes at 500 cells in 5 ml of Dulbecco's modified Eagle's medium per dish (Hansen and Stern 1983), and 24 h later the test compounds were added; the dishes were incu-

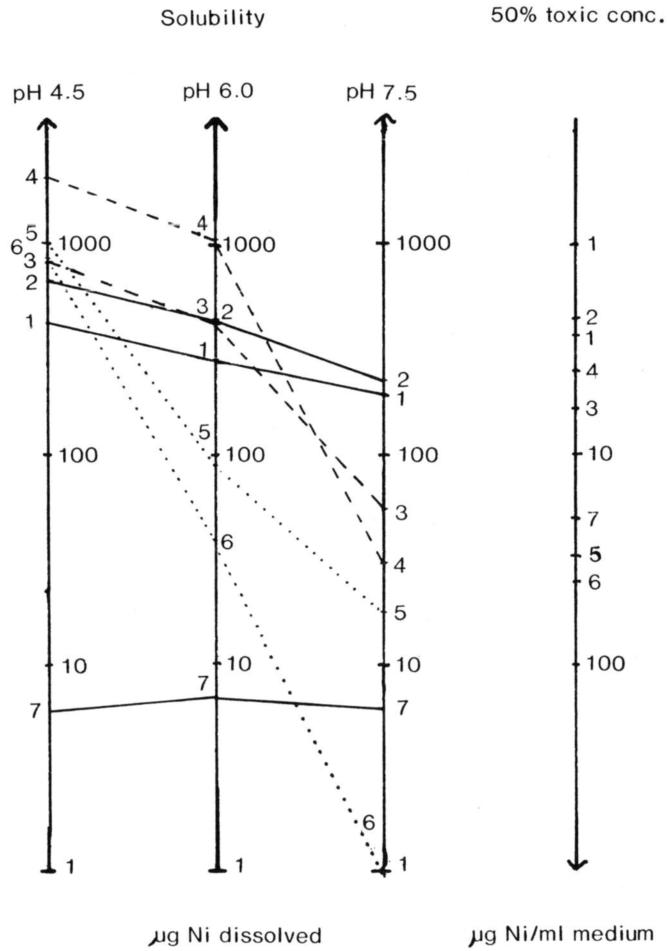

FIG. 1. Solubility in serum of nickel compounds at pH 4.5, 6.0 and 7.5 (mean of 2). Total amount of dissolved Ni after 12h of incubation at 37°C. 50% toxic concentrations in the BHK-21 assay (mean of 2 exp.). Compounds: 1) Black NiO (BDH Chem.), 2) Black NiO, 3) Ni metal, 4) Ni metal, 5) NiFe alloy, 6) NiFe alloy, 7) Green NiO; 2)-7) from INCO Ltd. Particle size: 90% smaller than 3 µm (except no. 6).

bated for 8 days at 37°C, pH 7.2-7.3, then the colonies were stained with crystal violet and the relative survival determined by counting the number of colonies in each dish. Each nickel compound was tested in 2 separate experiments in 3 concentrations with 5 dishes for each test point.

RESULTS AND DISCUSSION

All compounds at all three pH levels showed a time-dependent increase in the amount of dissolved nickel. The maximum amount of nickel which could be dissolved depended on the compound and on the pH level. All of the compounds were more soluble at acidic pH and the rate of dissolution increased with decreasing pH level.

The amount of nickel dissolved after 12h of incubation is chosen as suitable for illustrating solubility in serum at different pH levels (Fig. 1). The rank of solubility changes with increasing pH; e.g. the solubility of black NiO remains within the same decade whereas the solubility of metallic Ni decreases almost two decades going from pH 4.5 to pH 7.5.

At pH 6.0 the solubility of the compounds no. 1-4 lies within one decade corresponding to their 50% toxic concentrations which are also found within one decade. At pH 7.5 the solubility of metallic Ni is much lower than the solubility of black NiO. These findings suggest that the amount of available nickel depends a.o. on the solubility at an acidic pH found in the intracellular vacuoles.

ACKNOWLEDGEMENTS

The programming of the AAS-analysis by Britta Pedersen Ph.D. and the technical assistance of Anja Francis Madsen are gratefully acknowledged.

REFERENCES

Abbracchio MP, Simmons-Hansen J, Costa M (1982). Journal of Toxicology and Environmental Health, 9, 663-676.

Costa M, Simmons-Hansen J, Bedrossian CWM et al. (1981). Cancer Research, 41, 2867-2876.

Hansen K, Stern RM (1983). Environmental Health Perspectives, 51, 223-226.

Jensen MS, Bainten DF (1973). J of Cell Biology, 56, 379-388.

Kuehn K, Fraser CB, Sunderman Jr. FW (1982). Carcinogenesis, 3, 321-326.

Kuehn K, Sunderman Jr FW (1982). Journal of Inorganic Biochemistry, 17, 29-39.

461633

AN *IN VITRO* GENOTOXICITY ASSAY TO COMPLEMENT *IN VIVO* INHALATION STUDIES AND FOR OCCUPATIONAL MONITORING OF METALLIC AEROSOLS AND WELDING FUMES

RICHARD M. STERN* KAREN HANSEN** and EBBE THOMSEN***

*The Danish Welding Institute, DK-2600 Glostrup
**The Danish National Institute for Occupational Health, DK-2900 Hellerup
***Oldvejen 7, Røgerup, DK-4050 Skibby Denmark

INTRODUCTION

The control of occupational exposures is traditionally performed by hygiene monitoring with membrane filters. For the case of nickel, gravitational measurement of total dust is not sufficient as speciation is required: TLV's are, typically, 1.0 mg/m^3 and 0.1 mg/m^3 for insoluble and soluble fractions. Hansen and Stern (1984a) have demonstrated in the BHK-21 mammalian cell assay that a number of nickel aerosols including NiO and the fumes from welding using Ni wire, collected by traditional means, exhibit equal transformation potency at the 50% toxic dose, which varies widely with material.

Stern et al. (1984) have recently demonstrated that occupational exposures may contain short lived biologically reactive species which are not retained on membrane filter samples (e.g. Cr(VI)) and have described a method of collection in a microimpinger, which results in the fixation of the active species. Speciation into soluble and insoluble fractions in the impinger is not the same as that found for samples collected on membrane filters. It is therefore of interest to investigate the microimpinger as a tool for the collection and bioassay of metallic nickel aerosols by using a collection medium compatible with an *in vitro* bioassay system.

MATERIALS AND METHODS

Ni aerosol is produced by the metal inert gas (MIG) welding of a pure Ni filler wire (Inco 61, 1.2 mm, 16 V, 185 A or 0.8 mm, 18 V, 135 A) using argon-1% oxygen gas in a spray-arc configuration, downhand, at 6 sec/cm on sandblasted mild steel plate. Fume is collected via a 20 cmØ column, fitted with a cowl and a mixing disk. Impingers are placed 1.5 m above the welding arc. Ventilation is provided by two vacuum cleaner fans giving 1.5 m/sec updraft at 43 l/sec. The fume is a nonstoichiometric Ni:Ni-oxide mixture (60% Ni) (MMD = about 0.2µ).

Total fume is collected via a paper filter (Watman 41) placed at the top of the collection column. With the filter in place, after collection of 200-400 mg fume, the flow rate is of the order of 4 l/sec.

The impinger is a 250 ml flask with an extended lower section (15 cm long, 2 cm diameter) in which the inlet tube ends, in a capillary, 2-4 cm above the bottom. The impinger is filled with 30 ml of growth medium. A foam collector after the impinger protects the pump. With one 0.4 mmØ jet, the flow rate is 1.5 l/min: the impinger collects about 3-6 mg fume, at 60-70% efficiency, in 18 min. Multiple jets will increase the flow rate.

The BHK-21 mammalian cell in vitro transformation assay, modified for use with particulates, has been described in detail elsewhere (Hansen and Stern 1984a). For purposes of these present studies only toxicity was measured as it has been shown, for the case of Ni, to be directly related to transformation potential. To each 60 mm dish was seeded 500 baby hamster kidney cells (BHK-21 cell line) suspended in 5 ml Dulbecco's modified Eagle's medium. After 24 hours the test compounds were added and the dishes incubated for 8 days at 37°C (pH 7.2-7.3). The colonies were then stained with crystal violet and the relative survival determined by counting the number of colonies in each dish. Toxicity is measured as the average of 5 dishes at each concentration of test substance.

After collection, makeup medium was added to a total volume of 30 ml to replace that lost due to evaporation and spillover due to foaming, and the sample diluted to give at least three concentrations in the range of 10-90% toxicity. Two samples are prepared at each concentration, one for total analysis and one, centrifuged for 20 min at 4,000 rpm, to permit toxicity measurement of the soluble and insoluble fractions, separately. All samples were prepared and added to the bioassay within 3 hours of collection. An additional set of samples were made up from fume collected on the total paper filter at the top of the column, and from black NiO (MMD 1-3µ, 82%Ni). A soluble Ni sample was also prepared from the paper collected welding fume by incubation in medium for 24 hours. Nickel acetate (Merck 12278) was prepared by dissolution in water, as an example of a soluble Ni compound. Ni content is measured, before dilution, in each sample by AAS.

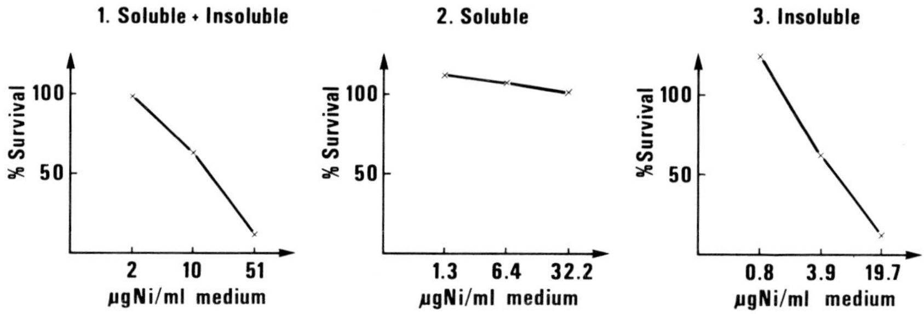

Fig. 1. Toxicity of Ni welding fume collected by impinger.

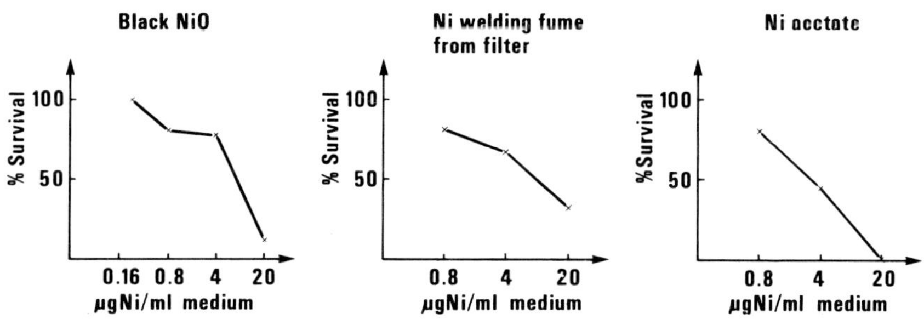

Fig. 2. Toxicity of soluble and insoluble Ni compounds

RESULTS

The results of toxicity vs dose experiments for Ni welding fume are shown in Fig. 1: 1) for the entire fume, 2) for the medium soluble fraction, 3) for the insoluble fraction, all from the same dilution series. The toxicity vs. Ni dose for the black NiO, Ni welding fume from filter, and nickel acetate are shown in Fig. 2 (average of 5 plates), and summarized in Table 1.

TABLE 1. 50% toxic concentrations in BHK-21 assay (μg Ni per ml medium)

	Total	Soluble	Insoluble
Ni fume(impinger)	14-15	(>32)	5-13
Ni fume(filter)	12	-	(∿11)
Black NiO	2.5-8	-	(2.5-8)
Ni acetate	-	3.5	-

DISCUSSION AND CONCLUSION

The insoluble fraction of Ni welding fume collected in an impinger filled with the growth medium for the BHK bioassay (pH 7.3) is totally responsible for the toxic effect of the entire fume: 50% toxic concentrations range from 5-12 μg Ni/ml medium. The toxicity is approximately the same as that of fume collected on paper filters, and of black NiO.

Although nickel acetate is 50% toxic at 3-4 μg Ni/ml medium, the growth-medium soluble fraction of fume collected in the impinger exhibits no toxicity up to concentration of 32 μg Ni/ml medium. The fume solubility in the medium filled impinger is 30-60% while that of paper collected fume is 5-10% at most. Apparently the enhanced solubility of the fume in medium is associated with the formation of organic ligands which effectively sequester the Ni in a non-bioavailable form. The potency or availability of the insoluble particulates is unaffected. A similar effect has been observed for Cr(VI) from stainless steel welding fumes collected and tested in the same way with BHK cells (Hansen and Stern 1984b). Thus for Ni and Cr(VI) metallic aerosols it is the extracellularly insoluble species which, after phagocytosis, mainly contribute to the activity of all but the most soluble substances tested in this type of bioassay.

Speciation and biological activity of metallic aerosols collected in an impinger yield different results from those collected on a membrane filter. If the impinger incorporates pathways and compartments similar to those found in the respiratory tract, it might become a collection and assay tool suitable for estimation of delivered dose per unit inhalation exposure and hence be relevant to in vivo toxicity testing and risk assessment.

ACKNOWLEDGEMENT

Research supported in part by CEC (Luxemburg) 4th ECSC medical research Program Contract 7248/22/016.

REFERENCES

Hansen K. and Stern R.M. (1984a) in Davis W. and Heseltine E. (Eds.), Nickel in the Human Environment, IARC, Lyon, in Press.

Hansen K. and Stern R.M. (1984b), submitted to Journal of Applied Toxicology.

Stern R.M. et al. (1984) Toxicological and Environmental Chemistry, 8, 95-108.

PULUMONARY DEPOSITION AND CLEARANCE OF INHALED NICKEL OXIDE AEROSOL

YASUSHI KODAMA, ISAMU TANAKA, KOJI MATSUNO, SHIGEKO ISHIMATSU, AKIO HORIE AND KENZABURO TSUCHIYA
School of Medicine,
University of Occupational and Environmental Health,
Yahatanishi, Kitakyushu 807, Japan

INTRODUCTION

This report describes the results of experimental studies in which rats were exposed to NiO aerosol by inhalation. The exposure conditions and aerosol properties were monitored to provide both physical and chemical data. The burden of NiO was measured by chemical analysis and then used to determine the effective deposition fractions.

EXPERIMENTAL

The exposure system and the exposure chamber have been fully reported in our previous work (Tanaka et al.,1982).

Respirable aerosols of NiO were generated and transported to exposure chambers. The NiO aerosol concentration in the exposure chamber was monitored continuously by a light scattering method and was measured gravimetrically at daily intervals by the isokinetic suction of air through a glass filter. The glass filters were weighed before and after the measurements. The size distribution of the aerosol was determined by using an Andersen sampler.

The rodents used in this study were Wistar male rats. The rats were randomly divided into controls and test groups. In all runs, the rats were exposed to NiO aerosols daily from 6 to 7 hours, 5 days per week for a maximum period of 3 months. Controls were exposed to clean air in adjacent, identical chambers under similar flow, temperature and humidity conditions. The six experiments are summarized in Table 1. In Run No.1,2,4,5 and 6, rats were sacrificed for analysis of nickel in the visceral organs just after the termination of exposure. In Run No.3, rats were exposed for 140 hours and kept for a one-year clearance period before sacrifice. The control rats were also sacrificed at the same time.

Rats were sacrificed and their visceral organs were removed for determination of nickel contents. The organs were weighed and then treated with a mixture of nitric acid and sulfuric acid until completely digested. The nickel amounts was measured by a flameless atomic absorption spectrophotometer (Hitachi Model 170-70).

RESULTS AND DISCUSSION

Body weight

The gains in body weight of the exposed and control groups were not significantly different.

Nickel contents in rat organs

The nickel concentrations in the lungs of the exposed rats for each run were much higher than those of the controls, but in other organs, there were no significant differences of nickel concentration between them.

Lung burdens and calculation of deposition

Macroscopically, the lungs of rats exposed to NiO in each run were gray in color in contrast to the normal pink color of the lungs obtained from the control rats.

Table 2 shows the NiO contents in rat lungs. The lung burden of NiO was determined from the measured nickel amount in the lungs.

Fig. 1 shows the dose-response relationship between the exposure level (Exposure concentration X Total exposure time) and the lung burden based on Table 2. The lung burden

Table 1 NiO inhalation exposure experimental design

Run No.	1	2		3		4	5	6	
Exposure Level	H	L	H	L	H	SH	H	L	H
Exposure period	Two-months	One-month		One-month		One-month	One-month	Three-months	
time of sacrifice	Just after two-months exposure	Just after one-month exposure		After one-year clearance period		Just afrer one-month exposure	Just after one-month exposure	Just after three-months exposure	
NiO concentration (mg/m^3)	6.5 ± 3.5	1.4 ± 0.5	7.0 ± 3.9	0.6 ± 0.3	8.0 ± 1.8	70 ± 21	6.9 ± 1.2	0.4 ± 0.1	1.1 ± 0.2
Total exposure time (hrs)	216	147		140		140	126	448	
NiO-size: Aerodynamic median diameter (μm)	1.2	1.2		1.2		4.0	4.0	0.7	0.6
NiO-size: Geometric standard deviation	2.2	2.2		2.2		2.0	2.0	1.5	1.5
NiO aerosol feed system	Indirect* (Buffer tank)	Indirect* (Buffer tank)		Indirect* (Buffer tank)		Direct** (Fluidized bed)	Direct** (fluidized bed)	Direct (Fluidized bed & Cyclone)**	Direct (Fluidized bed & Cyclone)**

* Tanaka et al., 1982 ** Tanaka and Akiyama , 1984

increased with the increasing of the exposure level depending on the aerodynamic mass median diameter.

In calculating the deposited fraction of NiO in the lungs, the average body weights of the animals during the respective exposure periods were used to calculate the expected minute volumes based on the Guyton's relationship (Guyton, 1947). The total volume of air inhaled during the exposure periods was multiplied by the average NiO concentration to yield the total amounts of inhaled NiO. Then the apparent deposition fraction was obtained for each rat as shown in Table 2.

Fig. 2 shows the effect of aerodynamic mass median diameter on the deposition fractions of inhaled NiO. The deposition fraction significantly decreased with inceasing of the

Table 2 Deposition of inhaled NiO in lung

Run No. & Exposure level	N	NiO concentration (mg/m^3)	NiO size (µm)	Total exposure time (hrs)	Estimated inhaled NiO* (µg±SD)	Measured lung burden NiO (µg±SD)	Apparent deposition fraction (%±SD)
1-H	3	6.5	1.2	216	16170± 530	2350±470	14.5±2.6
2-L	5	1.4	1.2	147	1800± 90	360± 60	19.8±3.0
2-H	5	7.0	1.2		8950± 400	1300± 40	14.5±0.5
4-SH	5	70	4.0	140	92400±3450	2130±830	2.3±0.9
5-H	12	6.9	4.0	126	5650± 280	220± 50	3.9±0.8
6-L	5	0.4	0.7	448	1840± 70	330± 40	18.0±1.6
6-H	4	1.1	0.6	448	5100± 100	1200± 70	23.4±1.0

* Estimated inhaled NiO (µg) = C x T x V

C : Exposure concentration of NiO ($µg/m^3$)

T : Total exposure time (hrs)

V : Breathing rate (m^3/hr) ; calculated by Guyton's relationship between V and body weight [wt]

$$V = 2.10 \times (\text{wt in Gram})^{3/4}$$

** Apparent deposition fraction (%) = $\left(\frac{\text{Measured lung burden NiO (µg)}}{\text{Estimated inhaled NiO (µg)}}\right) \times 100$

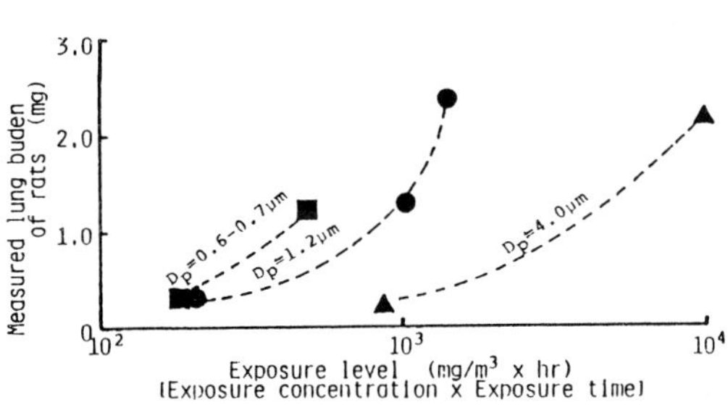

Fig. 1 Dose-response relationship between exposure level and lung burden.

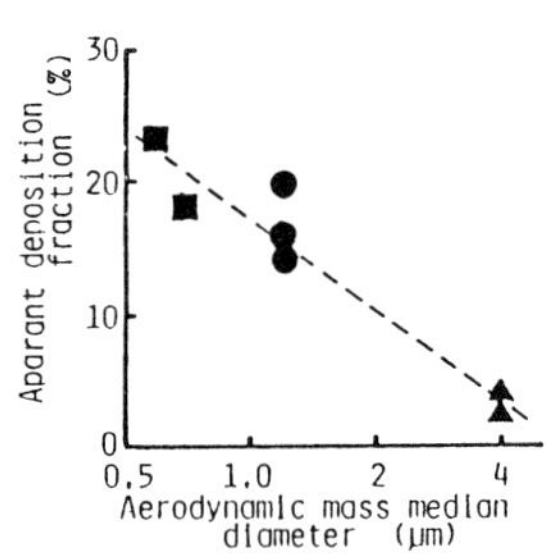

Fig. 2 Effect of NiO diameter on deposition.

diameter and slightly decreased with increasing of the exposure concentration. But there was not the effect of the exposure time as seen from Run No.1 and 2-H.

Clearance of nickel oxide in rat lung

The clearance of NiO deposited in rat lung is also important. In this study, the clearance rate was estimated as follows. As the experimental conditions in Run No.3-H and No.3-L seemed to be similar to those in Run No.2-H and 2-L, the apparent deposition fractions in Run No.3-H and No.3-L were taken from the results of Table 2 to be 14.5% and 19.8% respectively. The lung burden just after the termination of exposure for Run No.3 was calculated by multiplying the apparent deposition fraction by the total amount of inhaled NiO as shown Table 3. The clearance rate was obtained from the difference of the lung burden between just after the termination of exposure (estimated value) and after one-year clearance period (experimental value).

In this study, the clearance rate is estimated to be about 100 µg per year. The clearance is an important factor in health effects and should be monitored continuously.

Table 3 Clearance rate of inhaled NiO

Run No. & Exposure level	N	NiO Concentration (mg/m^3)	NiO size (µm)	Lung burden NiO (µg) Just after one-month exposure*	Lung burden NiO (µg) After one-year clearance period	Clearance rate NiO (µg/year)
3-L	5	0.6	1.2	140	50 ± 20	90
3-H	5	8.0	1.2	1370	1250 ± 480	120

* Estimated values

Lung burden NiO just after one-month exposure (µg) = Inhaled NiO X Apparent deposition franction

ACKNOWLEDGEMENTS

We wish to thank Mr. Onizuka of the Animal Research Center and Miss Ishimatsu who associated in sampling.

REFERENCES

Guyton A.C.,Am. J. Physiol. 150, 70 (1947)

Tanaka I.,Hayashi H.,Horie A.,Kodama Y.,Akiyama T. and Tuchiya K.,Journal of UOEH 5, 423 (1982)

Tanaka I. and Akiyama T.,Annals of Occupational Hygiene, 28, 157 (1984)

COMPARATIVE TOXICITY OF NICKEL SALTS TO THE LUNG

JANET M. BENSON*, ROGENE F. HENDERSON*,
ROGER O. McCLELLAN* and A. H. REBAR**
*Lovelace Inhalation Toxicology Research Institute,
Albuquerque, New Mexico, 87185 USA
**Department of Veterinary Microbiology, Pathology,
and Public Health, School of Veterinary Medicine,
Purdue University, W. Lafayette, Indiana 47907 USA

INTRODUCTION

Epidemiological studies of nickel refinery workers in Wales, Canada, Norway and Russia have documented increased incidence of lung and nasal cancers over that in control populations. Nickel compounds are considered as the responsible carcinogenic agents. To provide information on the toxicologic potential of inhaled nickel compounds, the Inhalation Toxicology Research Institute, in collaboration with the National Toxicology Program, will be conducting studies to compare the relative inhalation toxicities of three nickel salts (nickel oxide, nickel sulfate and nickel subsulfide) in Fischer-344 rats and B6C3F1 mice. To complement these studies, we have investigated the acute toxic effects of nickel oxide, nickel chloride, nickel sulfate and nickel subsulfide on the rat lung *in vivo* and on canine alveolar macrophages *in vitro*.

METHODS

Nickel oxide (NiO), nickel sulfate ($NiSO_4 \cdot 6H_2O$) and α-nickel subsulfide (Ni_3S_2) were obtained from the National Toxicology Program. Nickel chloride ($NiCl_2$) was purchased from Mallinkrodt Chemical Co., St. Louis, MO.

Groups of 48 (24 males and 24 females) laboratory-reared Fischer-344/Lov rats from the Institute's colony were administered a single nickel compound. Twelve animals within each group were administered either 0.01, 0.10, or 1.0 μmol nickel and twelve animals served as controls. Compounds were dissolved or suspended in physiological saline (0.15 M) containing 0.2% gelatin. Rats were anesthetized using halothane and intubated intratracheally as described by Mauderly (1977). Rats were sacrificed at one or 7 days following nickel administration and the heart-lung block was removed. The left lobe was fixed by intratracheal instillation of 10% neutral buffered formalin. Tissues were imbedded in paraffin, sectioned at 5 μm and stained with hematoxylin and eosin. The right lung was lavaged twice with physiological saline to sample the epithelial lining fluid. The bronchoalveolar lavage fluid obtained from each animal was analyzed as described by Henderson et al. (1979) for biochemical indicators of pulmonary cytotoxicity (extracellular lactate dehydrogenase, LDH) and for evidence of increased phagocytic activity (beta glucuronidase, BG). Total protein (TP) content of lavage fluid was quantitated to determine the extent of vascular leakage. To estimate the extent of inflammatory response, nucleated cell (NC) counts were made of cells recovered in the fluid. Mean values were calculated for all parameters evaluated for each exposure group and statistical significance was evaluated using the Student's t test with $p < .05$.

Macrophages for the in vitro studies were obtained by bronchoalveolar lavage of anesthetized beagle dogs. Nickel compounds were tested over a concentration range of 10 to 300 μg compound/ml. Cell viability (trypan blue exclusion) was determined at the end of a 20 h incubation period.

RESULTS

Minimal changes in LDH, BG, or TP content of alveolar lavage fluid were observed at one day post exposure. At seven days post exposure, the amount of LDH released into airways of rats administered $NiCl_2$, $NiSO_4$, and Ni_3S_2 increased in a dose-dependent manner. No increase in the amount of LDH in airways of animals exposed to any dose of NiO was observed. A dose-dependent increase in the amount of BG present in airways occurred in rats exposed to $NiCl_2$, $NiSO_4$, and Ni_3S_2. Exposure to all four Ni compounds resulted in an increase in the amount of TP in the airways. However, the response was not dose-dependent. The total numbers of nucleated cells (NC) recovered from the airways were significantly increased in rats exposed to 0.1 and 1.0 μmole Ni as $NiSO_4$ and Ni_3S_2, and 1.0 μmole Ni as $NiCl_2$. Statistically significant changes in the above parameters are summarized in Table 1.

Histologically, lesions from exposed animals were similar in character for all treatment regimens but varied in degree of severity. The principal alteration was multifocal alveolitis. Lungs from all animals were assigned a lesion index score on a scale of increasing severity from 0 to 4. Mean lesion scores and standard errors were computed for all groups. In animals sacrificed at one day post exposure,

TABLE 1. Summary of changes observed in lungs at 7 days post exposure

Compound	LDH*	BG*	TP*	NC*	Lesion Score**
$NiCl_2$					
0.1 μmol	126	NS	149	NS	2.5
1.0 μmol	253	320	140	172	2.3
$NiSO_4$					
0.1 μmol	143	NS	142	210	1.8
1.0 μmol	212	NS	104	188	2.5
Ni_3S_2					
0.1 μmol	113	NS	663	165	NS
1.0 μmol	346	221	521	247	3.0

Statistical significance was evaluated using the Student's *t* test with $P < 0.05$ in all cases.

*LDH, lactate dehydrogenase; BG, beta glucuronidase; TP, total protein; NC, nucleated cells. Data represent the per cent of mean control response.
NS, mean values not significantly different from control.

**Data represent the mean lesion scores evaluated as described in the text.

lesions were sporadic and mild and were most consistently present in the $NiSO_4$ and Ni_3S_2 highest dose groups.
In animals killed seven days post-exposure, there were no lesions in the NiO exposed animals at any dose level. Significant lesions were seen in animals exposed to the medium and high dose levels of $NiCl_2$ and $NiSO_4$, and the high dose of Ni_3S_2 (Table 1).

Median lethal concentrations of NiO, $NiCl_2$, $NiSO_4$, and Ni_3S_2 for canine alveolar macrophages were 225, 89, 92 and 6.0 μg compound/ml incubation medium, respectively.

DISCUSSION

Exposure of Fischer-344 rats to $NiCl_2$, $NiSO_4$, or Ni_3S_2 resulted in damage to or death of cells within the airways of the lung, in increased phagocytic activity within the lung, and in an inflammatory response. In most cases, damage was most severe in animals exposed to the highest dose level of Ni_3S_2. The morphological observations of mild to moderate alveolitis with multifocal involvement are consistent with the increases in LDH, BG, TP, and NC content of bronchoalveolar lavage fluid in animals exposed to $NiCl_2$, $NiSO_4$, and Ni_3S_2. Based on results of both the in vivo and in vitro studies, the compounds could be ranked in toxicity as follows: $Ni_3S_2 > NiCl_2 \sim NiSO_4 > NiO$.

The results obtained following exposure of rats to $NiCl_2$, $NiSO_4$, and Ni_3S_2 are similar to those reported by Henderson et al. (1979) for exposure of Syrian hamsters to $NiCl_2$. One major difference is that no morphological changes were observed in lungs of hamsters exposed to $NiCl_2$, possibly indicating that hamsters may be more resistant than rats to the effects of Ni compounds administered directly to the lung.

CONCLUDING REMARK

We have investigated the acute response of lung to administration of four nickel compounds. Results obtained will be used to gain a better understanding of the mechanisms leading to the development of late occurring health effects following chronic inhalation exposures of rodents to NiO, $NiSO_4$, and Ni_3S_2.

ACKNOWLEDGMENTS

This research was performed under U. S. Department of Energy Contract No. DE-AC04-76EV01013 in facilities fully accredited by the American Association for Laboratory Animal Care. The assistance of our professional and technical colleagues, especially Drs. R. K. Jones and J. A. Pickrell, is gratefully acknowledged.

REFERENCES

Henderson, R.F., Rebar, A.H. and DeNicola, D.B. (1979) Toxicology and Applied Pharmacology, 51, 129-135.

Mauderly, J. L. (1977) Laboratory Animal Science, 27, 255-261.

ALVEOLAR PROTEINOSIS INDUCED IN RATS BY LONG-TERM INHALATION OF NICKEL OXIDE

SHINJI TAKENAKA*, DIETER HOCHRAINER** and HUBERT OLDIGES**
Fraunhofer-Institut für Toxikologie und Aerosolforschung *Stadtfelddamm 35, 3000 Hannover **5948 Schmallenberg-Grafschaft, FRG

INTRODUCTION

Toxicity of nickel has been documented in reports on personnel occupationally exposed to nickel as well as in studies using experimental animals. Primarily, nickel dust, nickel subsulfide and nickel oxide, known to be insoluble in water, and nickel carbonyl vapour are suspected to be carcinogenic to the respiratory system.

At present, there is little information available concerning the chronic effects of inhaled nickel oxide (NiO). Therefore, we performed a long-term inhalation experiment to study the pulmonary responses of rats to low-level concentrations of this compound.

DETAIL

The inhalation was performed according to a technique previously reported by Prigge (1978) and Weischer et al. (1980). The exposure took place in 225-liter inhalation chambers comprising 2 wire mesh cages each containing 10 rats. We generated the aerosol by atomization of an aqueous Ni-acetate solution and subsequent pyrolysis of the particles.

100 male Wistar rats were used. They were approximately 8 weeks old and weighed 179-185 g at the beginning of the experiment. 40 and 20 animals were continuously (23 hr/d; 7 d/wk) exposed to NiO aerosols with nominal Ni-concentrations of 60 and 200 $\mu g/m^3$, respectively. The remaining 40 animals served as the control group. Both experimental and control animals received water ad libitum and a pellet diet. After the inhalation period, all rats were housed singly in plastic cages under conventional laboratory conditions. During the experiment, the animals were weighed every 2 to 4 months. Dead animals were autopsied immediately after they had been detected. The rats surviving 31 months after the beginning of the experiment were sacrificed by exsanguination under pentobarbital anesthesia for histopathology.

During the first 10 months of the inhalation period, there were no marked differences in the development of body weight among the exposure groups and the control group. After 13 months of exposure, both exposed groups had lost weight, which kept on decreasing until the end of the observation period. The control group started losing weight 24 months after the beginning of the experiment.

The mean survival times of 87.7 ± 16.7 weeks for the group exposed to 60 µg/m³ and 95.8 ± 15.0 weeks for the group exposed to 200 µg/m³ were significantly different from the mean survival time of 125.2 ± 12.5 weeks for the control group. The mean values include the lifetime (135 weeks) of rats sacrificed when the experiment was terminated (Fig. 1).

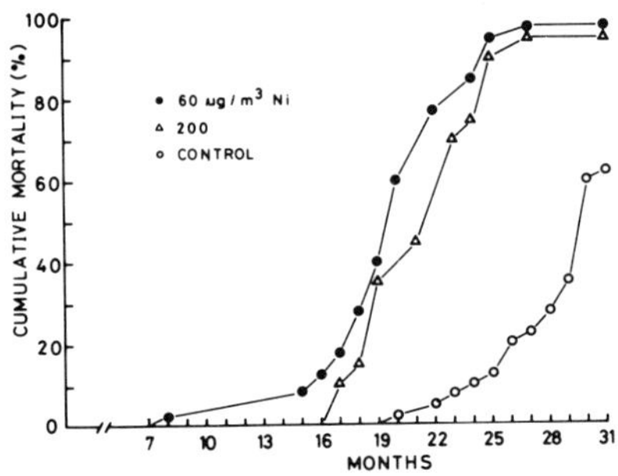

Fig. 1. Cumulative mortality during exposure and observation period

The major macroscopical observation made in the exposure groups was marked lung enlargement accompanied by congestion. Compared with the control group, both experimental groups showed an approximately sixfold increase in lung weight.

In every exposed animal examined histopathologically, the pulmonary alveolar lumina were filled with homogeneous, acidophilic and strongly PAS-positive material (Figs 2, 3). Alveolar walls had either normal architecture or sometimes appeared stretched. There were very few hyperplastic changes. In several cases the accumulation of foamy macrophages adjacent to the PAS-positive areas in the alveolar lumina was observed. For further characterisation of the intra-alveolar material we prepared samples for ultramicroscopy from lung tissue embedded in paraffin. The examined rats died at 21 months. As shown in Fig. 4, the substances consisted of lamellar meshwork or parallel lamellar pattern. Dense osmophilic bodies were accumulated in the lamellar meshwork. These findings are characteristic of alveolar proteinosis (Heppleston and Young, 1972). Two rats sacrificed at the end of the experiment had developed not only alveolar proteinosis with marked accumulation of macrophages but also focal septal fibrosis.

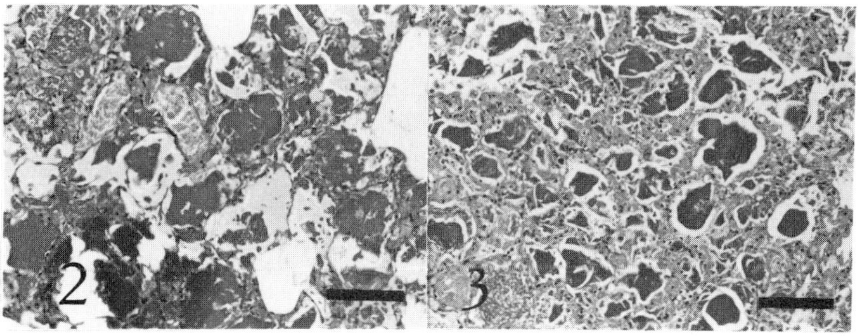

Fig. 2,3. Alveolar lumen filled with PAS-positive material in a rat from each of the groups exposed to 60 and 200 µg Ni/m^3. These rats died at 21 months after the beginning of the experiment. Bar = 100 µm.

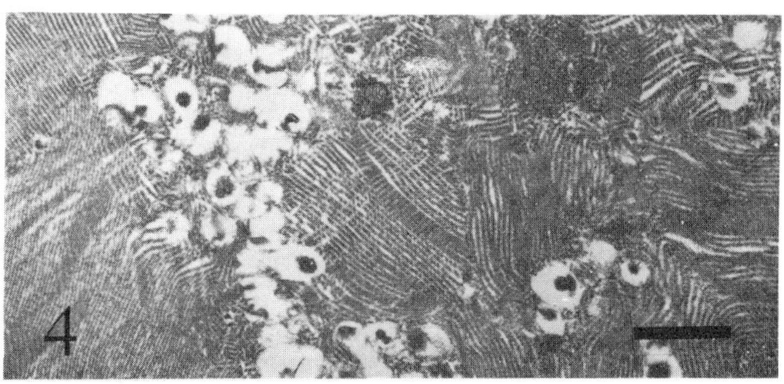

Fig. 4. Lamellar meshwork in the pulmonary alveolar lumen and dense osmophilic bodies in the lung of a rat exposed to 60 µg Ni/m^3, which died at 21 months.
Bar = 1 µm

DISCUSSION

A series of investigations recently performed by Johansson et al. (1983) and Camner et al. (1978) revealed pulmonary changes including alveolar proteinosis in rabbits exposed to metallic nickel dust over a maximum period of 6 months.

At present, the pathogenesis of this disease is still unsolved. The intra-alveolar substances are proposed to be at least in part derived from surfactant. Therefore, excessive accumulation of proteinaceous material appears to be dependent on the capability of clearance functions for surfactant, which involve 1. alveolar cells 2. centripetal movements and mucociliary escalator 3. lymphatics and 4. macrophages (Shingh et al., 1983). Long-term exposure to NiO-aerosols may inhibit one or more of these mechanisms, thus resulting in alveolar proteinosis.

At the end of this experiment, two surviving rats (one from each exposure group) showed not only alveolar proteinosis but also focal fibrosis. However, lung tumors were not observed in any rat. Since most of the animals exposed to 60 or 200 µg Ni/m³ either died at relatively early experimental stages or had severe proteinosis persisting toward the termination of the observation period, the further development of the affected lungs could not be examined. Long-term inhalation studies utilizing even lower concentrations of NiO may provide further information.

ACKNOWLEDGEMENTS

The authors wish to acknowledge the assistance of the staff of the ITA Grafschaft. This study was supported by the German Federal Environmental Agency and by the European Commission.

REFERENCES

Camner, P., Johansson, A. and Lundborg, M. (1978) Alveolar macrophages in rabbits exposed to nickel dust. Ultrastructural changes and effect on phagocytosis. Environ. Res. 16, 226-235.

Heppleston, A.G. and Young, A.E. (1972) Alveolar lipo-proteinosis: An ultrastructural comparison of the experimental and human forms. J. Pathol. 107, 107-117.

Johansson, A., Camner, P., Jarstrand, C. and Wiernik, A. (1983) Rabbit lungs after long-term exposure to low nickel dust concentrations. 2. Effects on morphology and function. Environ. Res. 30, 142-151.

Prigge, E. (1978) Early signs of oral and inhalative cadmium uptake in rats. Arch. Toxicol. 40, 231-247.

Shingh, G., Katyal, S.L., Bedrossian, W.M. and Rogers, R. (1983) Pulmonary alveolar proteinosis. Staining for surfactant apoprotein in alveolar proteinosis and in conditions simulating it. Chest 83, 82-86.

Weischer, C.H., Kördel, W. and Hochrainer, D. (1980) Effects of $NiCl_2$ and NiO in Wistar rats after oral uptake and inhalation exposure, respectively. Abl. Bakt. Hyg., 1. Abt. Orig. B. 172, 336-351.

INCREASED LIPID PEROXIDATION IN LIVER, KIDNEY, AND LUNG OF $NiCl_2$-TREATED RATS

F. WILLIAM SUNDERMAN JR., ODETTE ZAHARIA, ABUBAKR MARZOUK, AND MARILYN C. REID
Departments of Laboratory Medicine and Pharmacology, University of Connecticut Medical School, Farmington, Connecticut 06032, USA

INTRODUCTION

Lipid peroxidation is an autocatalytic free-radical process in which polyunsaturated fatty acids (e.g., arachidonic acid) in cell membranes undergo degradation by a chain-reaction to form lipid hydroperoxides, epoxy-fatty acids, alkanals, alkenals, 4-hydroxyalkenals, and aldehydes (e.g., malondialdehyde). Lipid peroxidation has been implicated as a mechanism for diverse types of tissue damage (e.g., irradiation, CCl_4, aging) (Recknagel et al., 1982). Recent studies suggest that lipid peroxidation may contribute to the toxicity of Cu, Cd, and Hg in rats (Dougherty and Hoekstra, 1982; Fukino et al., 1984; Sato et al., 1983; Stacy and Kappus, 1982). Fawade and Pawar (1983) reported increased lipid peroxidation in hepatic microsomes of $NiCl_2$-treated rats. The present study shows that lipid peroxidation is markedly increased in liver, kidney, and lung homogenates of $NiCl_2$-treated rats.

MATERIALS AND METHODS

The experimental animals were male Fischer-344 rats (180 - 250 g body weight) fasted 18 h before decapitation. Nickel chloride, dissolved in NaCl solution (0.14 mol/L), was administered by sc injection; control rats received sc injection of the NaCl vehicle. Lipid peroxidation in tissue homogenates was measured by the thiobarbituric acid reaction for malondialdehyde (MDA) precursors, as described by Uchiyama and Mihara (1978). Results are expressed as µmol MDA/g (dry wt) of tissue.

RESULTS

Lipid peroxidation was significantly increased in liver, kidney, and lung of rats killed 24 h after injection of $NiCl_2$ (0.75 mmol/kg) (Table 1). Maximum induction of lipid peroxidation in liver and kidney occurred at 24 to 48 hours after $NiCl_2$ injection (Table 2). Dose-effect relationships for $NiCl_2$ induction of lipid peroxidation are shown in Table 3. After a single injection of 0.125 mmol/kg, lipid peroxidation was significantly increased in rat kidney. After a single injection of 0.25 mmol/kg or three daily injections of 0.125 mmol/kg, lipid peroxidation was significantly increased in rat liver. For comparison, the acute (two-week) LD5, LD50, and LD95 dosages of $NiCl_2$ sc in male Fischer rats are 0.31, 0.42, and 0.56 mmol/kg, respectively. At sc dosages of $NiCl_2$ in the LD95 - LD100 range (0.6 to 1.0 mmol/kg), the median time between injection and death is 3.4 days (10th and 90th percentiles = 2 and 5 days, respectively) (Sunderman et al. 1983).

TABLE 1. Lipid peroxidation in tissues of control and $NiCl_2$-treated rats

Organ	MDA precursors (mean ± SD) (µmol/g dry wt) Controls (N = 6)	$NiCl_2$ (N = 7)
Liver	0.58 ± 0.20	2.00 ± 0.44[b]
Kidney	0.40 ± 0.03	0.69 ± 0.05[b]
Lung	0.15 ± 0.03	0.28 ± 0.08[b]
Brain	0.38 ± 0.05	0.39 ± 0.10
Heart	0.14 ± 0.06	0.13 ± 0.04
Spleen	0.32 ± 0.08	0.33 ± 0.10
Testis	0.47 ± 0.07	0.45 ± 0.09

[a] Injection of NaCl vehicle (controls) or $NiCl_2$ (0.75 mmol/kg, sc) 24 h before death.
[b] $P < 0.001$ versus controls.

TABLE 2. Time-course of lipid peroxidation in liver and kidney of $NiCl_2$-treated rats

Treatment	Hours after injection	No. of rats	MDA precursors (mean ± SD) (µmol/g dry wt) liver	kidney
Controls[a]	3 - 24	8	0.49 ± 0.15	0.38 ± 0.06
$NiCl_2$[b]	3	6	0.52 ± 0.19	0.34 ± 0.06
	9	5	0.76 ± 0.25[c]	0.64 ± 0.12[d]
	18	11	1.63 ± 0.32[d]	0.70 ± 0.11[d]
	24	13	2.23 ± 0.81[d]	0.71 ± 0.10[d]
	48	6	2.34 ± 0.58[d]	0.50 ± 0.07[d]

[a] NaCl vehicle, sc.
[b] $NiCl_2$, 0.75 mmol/kg, sc.
[c] $P < 0.05$ vs controls.
[d] $P < 0.001$ vs controls.

TABLE 3. Dose-effect study of lipid peroxidation in liver and kidney of $NiCl_2$-treated rats

$NiCl_2$ dose[a] (mmol/kg, sc)	No. of daily injections	No. of rats	MDA precursors (mean ± SD) (µmol/g dry wt) liver	kidney
0 (vehicle)	1	20	0.51 ± 0.21	0.39 ± 0.06
0.125	1	4	0.56 ± 0.19	0.48 ± 0.02[b]
0.25	1	5	0.78 ± 0.05[b]	0.48 ± 0.09
0.50	1	5	1.53 ± 0.54[b]	0.52 ± 0.05[b]
0.75	1	11	2.06 ± 0.74[b]	0.72 ± 0.10[b]
1.00	1	5	1.92 ± 0.27[b]	0.48 ± 0.08[b]
0.125	2	4	0.71 ± 0.21	0.59 ± 0.06[b]
0.125	3	4	0.75 ± 0.10[b]	0.69 ± 0.11[b]
0.125	4	5	0.73 ± 0.04[b]	0.50 ± 0.07[b]

[a] Injection, sc, daily, with death 24 h after the last injection.
[b] $P < 0.01$ versus vehicle controls.

DISCUSSION

This study implicates lipid peroxidation as a molecular mechanism for cell injury and death in acute $NiCl_2$ poisoning. The authors speculate that by forming highly reactive, genotoxic metabolites (e.g., 4-hydroxyalkenals), lipid peroxidation may cause DNA strandbreaks and DNA-protein crosslinks that could induce neoplastic transformation.

SUMMARY

Administration of $NiCl_2$ to rats enhances lipid peroxidation in liver, kidney, and lung, measured by the thiobarbituric acid reaction for malondialdehyde (MDA) precursors. After sc injection of $NiCl_2$ (0.75 mmol/kg), MDA precursors in liver and kidney become increased by 9 h and reach peak levels at 24-48 h. A dose-effect relationship for lipid peroxidation in liver and kidney is observed with $NiCl_2$ dosages ranging from 0.125 to 1.0 mmol/kg, sc.

REFERENCES

Dougherty J.J. and Hoekstra W.J. (1982) Proc. Soc. Exp. Biol. Med., 169, 201-208.
Fawade M.M. and Pawar, S.S. (1983) Ind. J. Exp. Biol., 21, 343-346.
Fukino H., Hirai M., Hsueh Y.M. and Yamane Y. (1984) Toxicol. Appl. Pharmacol., 73, 395-401.
Recknagel R.O., Glende E.A. Jr., Waller R.L. and Lowrey K. (1982) in Plaa G. and Hewitt W.R. (Eds.), Toxicology of the Liver, pp. 213-241, Raven Press, New York.
Sato M., Yamanoke K. and Nagai Y. (1983) Life Sci. 33, 903-908.
Sunderman F.W. Jr., Reid M.C., Bibeau L.M. and Linden J.V. (1983) Toxicol. Appl. Pharmacol. 68, 87-95.
Uchiyama M. and Mihara M. (1978) Anal. Biochem., 86, 271-278.

EFFECTS OF UNILATERAL INTRARENAL ADMINISTRATION OF NICKEL SUBSULFIDE TO RATS ON ERYTHROPOIETIN CONCENTRATIONS IN SERUM AND IN EXTRACTS OF BOTH KIDNEYS

SIDNEY M. HOPFER,* F. WILLIAM SUNDERMAN Jr.,* and EUGENE GOLDWASSER**
*Department of Laboratory Medicine, University of Connecticut Medical School, Farmington, CT 06032, USA
**Department of Biochemistry and Molecular Biology, University of Chicago Medical School, Chicago, IL 60637

INTRODUCTION

Pronounced erythrocytosis develops in rats within one month after intrarenal (ir) injection of nickel subsulfide, Ni_3S_2, (Jasmin and Solymoss 1975; Morse et al. 1977; Hopfer et al. 1980). Ni_3S_2-induced erythrocytosis is apparently mediated by increased renal production of erythropoietin (epo), since (a) epo activity is increased in serum of Ni_3S_2-treated rats, as determined by bioassay and by radioimmunoassay techniques (Solymoss and Jasmin 1978; Hopfer et al. 1979, 1984; Meagher 1980), (b) immunoreactive epo concentrations are markedly increased in kidney extracts of rats following bilateral ir injections of Ni_3S_2 (Hopfer et al. 1984), and (c) Ni_3S_2-induced erythrocytosis is prevented by protracted exposure of rats to hyperoxia or by ip infusion of anti-epo antiserum (Meagher, 1980).

Uncertainty exists whether unilateral ir administration of Ni_3S_2 to rats stimulates epo production in both kidneys or only in the injected kidney. Glomerular mesangial cells, which have been suggested as a site of renal epo production (Kurtz et al. 1982), become markedly hyperplastic in both kidneys of rats that receive unilateral ir injection of Ni_3S_2, suggesting that epo production is increased in both kidneys (McCully et al. 1982a,b). On the other hand, the erythrocytosis that develops in rats after unilateral ir injection of Ni_3S_2 is completely prevented by resection of the injected kidney and is unaffected by resection of the contralateral kidney, suggesting that epo production is increased only in the injected kidney (Sunderman et al. 1982). The present study was performed to resolve this controversy and to provide further evidence on the role of epo in the pathogenesis of Ni_3S_2-induced erythrocytosis.

MATERIALS AND METHODS

The animals were 12 male rats of the Fischer-344 strain, approximately 8 weeks old at the beginning of the experiment (body wt 250 ± 25 g). The rats were housed and fed as previously described (Hopfer et al. 1984). One-half of the rats received ir injection of Ni_3S_2 (10 mg/rat) in the right kidney. The remaining rats constituted the control group; these rats received ir injection of NaCl vehicle (0.14 mol/L, 0.2 ml/rat) in the right kidney. The injections were performed as described by Hopfer et al. (1984). Prior to the ir injection (day 0) and at 21 days post-injection, blood was collected from the tip of the tail for measurement of hematocrit (Strumia et al. 1954). On day 21, each rat was anesthetized with diethyl ether and exsanguinated by cardiac puncture. The blood tube was kept in an ice bath for 1 hour during clotting; serum was transferred to a chilled tube and frozen at -20°C until the time of assay. The abdomen was opened and the kidneys were removed without perfusion. The ureters, major blood vessels, and renal capsule were rapidly excised. The parenchyma of each kidney was minced with scissors and homogenized for 30 sec by use of a Polytron tissue homogenizer (setting = 2.8, Brinkman Instruments, Inc., Westbury, NY) in 5 ml of phosphate buffer (0.1 mol/L, pH 7.4, 0°C). The homogenate was centrifuged at 16,000 x G for 30 min at 4°C using a Sorvall model RC 2B centrifuge (Sherwood and Goldwasser, 1978). The supernatant kidney extract was frozen at -20°C until the time of assay. Epo-radioimmunoassay was performed by the method of Sherwood and Goldwasser (1979) using a rabbit antibody to human epo that cross-reacts with rat epo. The epo-radioimmunoassay was standardized by assaying rat sera with known epo concentrations, as determined by the plethoric mouse assay. Statistical comparisons of results in the control and Ni_3S_2-treated groups were performed by the Mann-Whitney U test (Siegel, 1956).

RESULTS

The effects of unilateral ir injection of Ni_3S_2 on blood hematocrit values and epo concentrations in serum and in kidney extracts are listed in Table 1. In Ni_3S_2-treated rats, blood hematocrit values and serum epo concentrations were significantly increased at 21 days after the ir injection, consistent with previous results (Hopfer et al. 1984). Epo concentrations were increased 3-fold in the injected kidneys of Ni_3S_2-treated rats and were unchanged in the contralateral, non-injected kidneys.

TABLE 1. Effects of unilateral injection of Ni_3S_2 on immunoreactive erythropoietin in rat serum and kidney extracts

Measurement	Days after injection[a]	Control rats (N=6)[b]	Ni_3S_2-treated rats (N=6)[b]
Blood hematocrit (%)	0	50 ± 2	50 ± 1
	21	50 ± 2	64 ± 4[c]
Serum epo (mU/mL)	21	12 ± 2	94 ± 81[c]
Epo in injected kidney (mU/g)	21	39 ± 20	118 ± 82[c]
Epo in non-injected kidney (mU/g)	21	30 ± 8	28 ± 5

[a] Ni_3S_2 (10 mg/rat) or NaCl vehicle (0.2 ml/rat) injected into the right kidney.
[b] Mean values ± S.D. [c] $P < 0.001$ versus controls.

DISCUSSION

This study shows that ir administration of Ni_3S_2 to rats increases epo production in the injected kidney, but not in the contralateral, non-injected kidney. This finding negates the hypothesis that release of a stimulatory substance (e.g., prostaglandin, renin, interferon) from the Ni_3S_2-injected kidney might induce epo production in both kidneys. The increased concentration of immunoreactive epo that occurred in the Ni_3S_2-treated kidney cannot be attributed to enhanced renal retention of serum epo during its excretion or catabolism, since the rates of renal uptake, degradation, and urinary elimination of circulating epo would be expected to be equal in both kidneys.

SUMMARY

Radioimmunoassay of erythropoietin (epo) was performed on serum samples and extracts of both kidneys from rats that received unilateral intrarenal injection of nickel subsulfide (Ni_3S_2, 10 mg/rat). Serum epo concentrations were significantly increased at 3 weeks after Ni_3S_2 injection (94 ± 81 mU/mL in 6 Ni_3S_2-treated rats vs 12 ± 2 mU/mL in 6 vehicle-treated control rats, $P < 0.001$). In extracts of the injected kidneys of rats killed 3 weeks after Ni_3S_2 injection, epo concentrations were increased 3-fold (118 ± 82 mU/g vs 39 ± 20 mU/g in vehicle controls, $P < 0.001$). Epo concentrations in extracts of the contralateral, non-injected kidneys were equal in the Ni_3S_2-treated and control rats (28 ± 5 mU/g in Ni_3S_2-treated rats vs 30 ± 8 mU/g in controls). These results indicate that ir administration of Ni_3S_2 to rats increases epo production in the injected kidney, but not in the contralateral, non-injected kidney. This finding excludes the possibility that increased epo concentration in kidney extracts of Ni_3S_2-treated rats might reflect renal retention of epo during its excretion or catabolism, since such an increase would be bilateral.

REFERENCES

Hopfer S.M., Sunderman F.W. Jr., Fredrickson T.N. and Morse E.E. (1979) Res. Commun. Chem. Pathol. Pharmacol., 23 155-170.

Hopfer S.M., Sunderman F.W. Jr., Morse E.E. and Fredrickson T.N. (1980) Ann. Clin. Lab. Sci. 10, 54-64.

Hopfer S.M., Sunderman F.W. Jr., Reid M.C. and Goldwasser E. (1984) Res. Commun. Chem. Pathol. Pharmacol. 43, 299-305.

Jasmin G. and Solymoss B. (1975) Proc. Soc. Exp. Biol. Med., 148, 774-776.

Kurtz A., Jelkmann W. and Bauer C. (1982) FEBS Lett. 137, 129-132.

McCully K.S., Rinehimer L.A., Gillies C.G., Hopfer S.M. and Sunderman F.W. Jr. (1982a) Virchows Arch (Pathol Anat) 394, 202-220.

McCully K.S., Sunderman F.W. Jr., Hopfer S.M., Kevorkian C.B. and Reid M.C. (1982b) Virchows Arch (Pathol Anat) 397, 251-259.

Meagher R.C. (1980) Mechanisms Underlying the Induction of Erythropoiesis in Rats by Intrarenal Implantation of Nickel Subsulfide (Ph.D. Thesis, New York Univ) pp 1-123, Univ. Microfilms Internat. Ann Arbor.

Morse E.E., Lee T-Y, Reiss R.F. and Sunderman F.W. Jr. (1977) Ann. Clin. Lab. Sci. 7, 17-24.

Sherwood J.B. and Goldwasser E. (1978) Endocrinol. 103, 866-870.

Sherwood J.B. and Goldwasser E. (1979) Blood, 54, 885-893.

Siegel S. (1956) Nonparametric Statistics for the Behavioral Sciences, pp 96-127, McGraw-Hill Co., New York.

Solymoss B. and Jasmin G. (1978) Exp. Hemat. 6, 43-47.

Strumia M.M., Sample A.B. and Hart E.D. (1954) Amer. J. Clin. Pathol. 24, 1016-1024.

Sunderman F.W. Jr., Hopfer S.M., Reid M.C., Shen S.K. and Kevorkian C.B. (1982) Yale J. Biol. Med. 55, 123-136.

FURTHER INSIGHTS INTO MECHANISMS OF NICKEL-INDUCED DNA DAMAGE: STUDIES WITH CULTURED RAT LIVER CELLS

SABINE H.H. SWIERENGA* and JACK R. McLEAN**
*Drug Toxicology Division and **Radiation Protection Bureau, Health and Welfare Canada, Ottawa K1A 0L2

INTRODUCTION

For toxicity testing in vitro with mammalian cells, endpoints such as DNA repair, mutagenesis, and morphological transformation, are considered to be cancer-related lesions. Numerous studies, predominantly with rodent fibroblasts, have demonstrated such genotoxicity of nickel compounds. In order to further investigate the nature of nickel-induced cellular lesions, we have used epithelial cells derived from normal rat liver (T51B; Swierenga et al., 1978) as a model to study genotoxic and cytotoxic effects of nickel compounds at concentrations leading to cancer-related endpoints.

NICKEL COMPOUNDS, CYTOTOXICITY AND MUTAGENICITY

In order to assess whether phagocytosis of insoluble particles of the potent in vivo carcinogen Ni_3S_2 contributes to in vitro genotoxicity, this compound was tested in 2 forms: (1) a freshly prepared aqueous suspension of washed particles, (2) an aqueous solution of Ni_3S_2 which was prepared by stirring for 3 days at 37°C, centrifugation, and filtration of the supernatant through a 0.2 µM Nalgene® filter. Ni(II) content was analysed using 4-(2-pyridylazo)-resorcinol as a chromogen. Soluble nickel chloride, which is non-carcinogenic in vivo, was also tested, as this compound has been reported as both positive (Amacher and Paillet, 1980) and negative (Painter and Howard, 1982) in in vitro assays. Cytotoxicities of these 3 forms of nickel as determined by survival of clonable cells differed markedly (Fig. 1). Nickel chloride was much less cytotoxic, in terms of concentration required to yield 50% survival (LC_{50}) and slope of survival curve, than the nickel sulfide compounds.

Mutations at the HGPRT locus were determined according to the method of Tong et al. (1980) with the selection agent 8-azaguanine. Unlike most carcinogens tested in this system, the nickel compounds were mutagenic only at very toxic concentrations (Table 1). All three forms of nickel tested were mutagenic but at very different effective concentrations. Thus neither insolubility nor phagocytosis were essential for mutagenicity; however, the presence of residual nickel due to phagocytosis was probably a factor in the enhanced response observed with particulate Ni_3S_2 (Table 1).

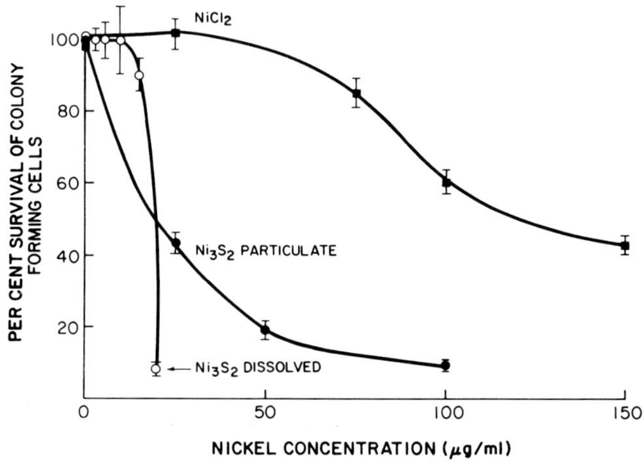

FIG. 1. Effect of various nickel compounds on colony-forming ability of rat liver cells. Cells were treated for 24 hr and replated at clonal seeding density.

DNA EFFECTS

The ability of nickel compounds to cause DNA strand breaks was assessed by fluorescent analysis of DNA unwinding, a sensitive technique described by Birnboim and Jevcak (1982). Strand breaks were only detected following treatment with mutagenic doses of Ni_3S_2 but these were associated with low survival and did not occur in a concentration-dependent manner (Table 2) suggesting that the observed breaks may have been related to cell death and not necessarily to mutagenic events. Breaks were not observed following treatment of resting (confluent) cells or cells treated with sub-toxic concentrations of nickel compounds.

All 3 nickel compounds severely suppressed DNA synthesis in exponentially growing cells (determined autoradiographically). However, only mutagenic (i.e., very toxic) concentrations of $NiCl_2$ and Ni_3S_2(dissolved) caused significant post-treatment suppression of DNA synthesis, an *in vitro* property of carcinogens (Painter and Howard, 1982). Particulate Ni_3S_2 caused irreversible post-treatment suppression of DNA synthesis at all concentrations, probably because residual particles could not be washed out (not shown).

TABLE 1. Effect of nickel compounds on survival and mutant frequency of T51B cells.

Treatment (24 hr)	Per cent survivors	No. of mutant colonies per 10^6 survivors
Untreated	100	69 ± 55
$NiCl_2$		
50 µg/ml	100	88 ± 25
100 µg/ml	70	116 ± 45
230 µg/ml	4.5	3148 ± 1283
Ni_3S_2 (particulate)		
5 µg/ml	100	278 ± 51
10 µg/ml	60	703 ± 83
50 µg/ml	20	1486 ± 241
Ni_3S_2 (dissolved)		
12.5 µg/ml	116	124 ± 68
15 µg/ml	92	198 ± 47
20 µg/ml	8.5	1504 ± 445

TABLE 2. Effect of Ni_3S_2 on survival and DNA strand breaks in T51B cells.

Treatment (18 hr)	Per cent survivors	Strand breaks per cell*
Cobalt-60 1 Gray	90	1200
Ni_3S_2 (dissolved)		
17.5 µg/ml	60	920
20 µg/ml	50	720
28 µg/ml	10	650

*Calibrated against Cobalt-60 gamma rays, 1 Gray is known to induce about 1200 breaks per genome.

DNA repair following nickel exposure was assessed in freshly isolated rat hepatocytes by measuring unscheduled DNA synthesis (UDS) autoradiographically according to the method of Williams (1978). Nickel, as both dissolved and particulate Ni_3S_2, 2-27 µg/ml, did not induce repair (not shown). Further, Ni_3S_2 inhibited repair of DNA damage induced by the alkylating agent methylmethanesulfonate, in a concentration-dependent manner. If nickel can prevent repair of damage caused by other carcinogens, enhanced cell death and genotoxicity should occur under these circumstances. This was observed in this study, and by others.

The failure of Ni_3S_2 to induce UDS raised the possibility that nickel could cause DNA damage resulting in "short patch" repair such as might be expected from production of free radicals, and difficult to detect in the UDS assay. We considered this to be unlikely, firstly because of nickel's known cross-linking activity that would result in long-patch repair, easily detectable by UDS and, secondly, because nickel compounds in aqueous media at pH 7.4, unlike radiation or Fe(II), were unable to activate oxygen to superoxide or hydroxyl radicals as determined with fluorescent (terephthalic acid) and colorimetric (nitroblue tetrazolium) indicators.

CYTOSKELETAL EFFECTS

In addition to mitotic abnormalities caused by interference of nickel with spindle proteins, we have also observed a marked effect of Ni_3S_2 on intermediate filament morphology of T51B cells, as determined by indirect immunofluorescence with specific monoclonal antibody to a 55K cytokeratin prepared from rat hepatocytes (generously provided by N. Marceau, University of Laval, Québec, Canada). A 24 hr exposure to mutagenic concentrations of Ni_3S_2 caused massive tangles of cytokeratin adjacent to the nucleus. These tangles increased in size with increasing duration of nickel exposure, and were observed in some of the cells to be irreversible and heritable. Similar lesions have been described previously by Borenfreund et al. (1980) during hepatocarcinogenesis in vivo. Indeed, along with the emerging importance of cytoskeletal proteins in mediation of growth controls and neoplastic transformation, other cancer-related effects of nickel compounds, such as inhibition and infidelity of DNA synthesis, could be explained through a direct effect on protein structure.

REFERENCES

Amacher D.E. and Paillet, S.C. (1980) Mutation Research 78, 279-288.

Birnboim H.C. and Jevcak, J.J. (1982) Cancer Research 41, 1889-1892.

Borenfreund E., Higgins P.J. and Bendich A. (1980) Annals of the New York Academy of Sciences 349, 357-371.

Painter R.B. and Howard, R. (1982) Mutation Research 92, 427-437.

Swierenga S.H.H., Whitfield J.F. and Karasaki, S. (1978) Proceedings of the National Academy of Sciences USA 75, 6069-6072.

Tong C. and Williams G.M. (1980) Mutation Research 74, 1-9.

Williams G.M. (1978) Cancer Letters 4, 69-75.

RELATIONSHIPS BETWEEN NICKEL INGESTION AND CALCIUM, MAGNESIUM AND GLUCOSE IN THE RAT.

NEY LOBATO RODRIGUES*, ETHELL BARBOSA NOVELLI* ICIAR BASAGOITI SATRUSTEGUI**, M.ISABEL SANCHEZ REUS** and BARTOLOME RIBAS OZONAS**.
Departments of Biochemistry: *Instituto Básico de Biología Médica é Agrícola, UNESP, Sao Paulo, Brasil **Consejo Superior de Investigaciones Científicas, Universidad Complutense de Madrid, Facultad de Farmacia, 28003 Madrid, Spain

INTRODUCTION

Calcium, phosphorus and magnesium are elements with structural and metabolic significance, principally as biochemical homeostatic modulators. There are interesting physiological and toxicological interactions between nickel and other bivalent cations(Anke et al., 1974; Sunderman Jr., 1977; Sanchez Reus et al., 1984). Low levels of magnesium favors calcium deposition, and hypermagnesemia after perinatal magnesium treatment induces hypocalcemia and also an increasing of parathormone levels.(Cruikshank et el., 1979).

Nickel deficiency induces weight inhibition, delay in sexual development and calcium, zinc and iron diminution in tissues (Anke et al., 1974; Kirchgessner and Schnegg, 1980). Nevertheless oral chronic nickel ingestion induces zinc and calcium incorporation into rat pancreas (Ribas et al., 1981), a fact which is probably connected with the increase of insulin secretion.

MATERIAL AND METHODS

Groups of 20 rats, males and females, of 210-160g, 20 days old at the beginning of the experiment and being the fourth generation given 200 μg Ni/ml water ($NiCl_2$) to drink, were sacrificed, blood extracted for Ca, P, Mg and glucose determinations by cardiac puncture. Liver and pancreas were dissected, cleaned, weighed, homogenized in 0.4 M buffer phosphate pH 7.4 (W:V, 1:2) and glucose determined with glucose-oxidase.
Groups of male Wistar rats of 240-270g weight were injected with 4 mg Ni/kg rat ($NiCl_2$) once per day for 6 days. Glucose tolerance curves were established after a single dose of intravenous 0.5 g glucose/kg body weight between 0 and 120 min,

24 hours after the last nickel dose. Mineral elements were determined by atomic absorption spectrophotometry.

RESULTS

Calcemia in nickel chloride-drinking male rats was diminished compared to control animals from 2.27±0.4 mM to 1.27±0.06 mM Ca, but in female rats with nickel ingestion it was increased from 1.75±0.19 to 2.15±0.1 mM Ca (Table 1). In female rats phosphorus was more increased than calcium. Blood Ca/P ratio in males was 0.7, which showed a diminution of calcium levels during nickel chloride ingestion, and in females a Ca/P had a quotient ratio of 0.9; nickel ingestion increased the plasma levels of magnesium in male and female rats as shown in Table 1. Phosphorus did not show any relevant variation in males; in females however the plasma levels increased (Table 1). Plasma glucose levels of nickel-drinking rats were diminished from 4.30±0.13 in control animals to 3.73 ±0.06 mM in treated animals, males and females. Nevertheless liver glucose of nickel treated animals was slightly increased in comparison with control ones, but significantly in the pancreas of both male and female rats from 0.96±0.04 to 1.33±0.15 mM and from 0.87±0.1 to 1.22±0.09 mM glucose, respectively.

TABLE 1. Calcium, phosphorus and magnesium concentrations in rat serum after oral nickel chloride ingestion (n = 10)

Animals	Calcium (mM)	Phosphorus (mM)	Magnesium (mM)
	(mean values ± S.D.)		
Control ♂	2.2 ± 0.40	2.0 ± 0.3	0.4 ± 0.05
Ni ingestion ♂	1.2 ± 0.06	1.8 ± 0.1	0.7 ± 0.03
Control ♀	1.7 ± 0.19	1.5 ± 0.2	0.7 ± 0.06
Ni ingestion ♀	2.1 ± 0.10	2.2 ± 0.1	1.4 ± 0.14

After intraperitoneal nickel chloride administration, the single dose of glucose was associated with a lowering of calcemia after 10 and 30 min (Fig 1), according to the theories of secretion and biosynthetic stimulation of insulin by glucose. As for calcium, zinc is also decreased in plasma of control rats, but in nickel-injected rats there is an increase in zinc concentration corresponding to insulin secretion and an increase in pancreatic zinc accumulation. Control rats show, as for calcium, a decrease in plasma zinc concentration which could signify the need of the pancreas for the synthesis of insulin, and change in the pancreas zinc homeostasis after the glucose stimulus. Nevertheless, the nickel-injected rats show an elevation of zinc (Fig 1), which could reflect the increase in insulin secretion and pancreatic incorporation of zinc.

DISCUSSION

Cruikshank et al. (1981) suggested that the hypermagnesemia induced through intravenous administration of magnesium could provoke hypocalcemia.This effect could be related to the inhibition of blood calcium levels by nickel ingestion in males and the consequent hypermagnesemia. It is known that oxalate intoxication leads to the same effects. Ribas et al. (1981) observed an incorporation of calcium and zinc into the pancreas of nickel drinking-rats corresponding to an increase of the physiological activity of the rat pancreas i.e. the insulin secretion, which could also explain the hypocalcemia and hypermagnesemia after nickel ingestion.

Some data expressed here are in agreement with findings of Rayssiguier et al. (1981) in rats. Magnesemia with nickel ingestion could also occur in connection with the inhibition of blood glucose as described by Rodrigues et al. (1983).

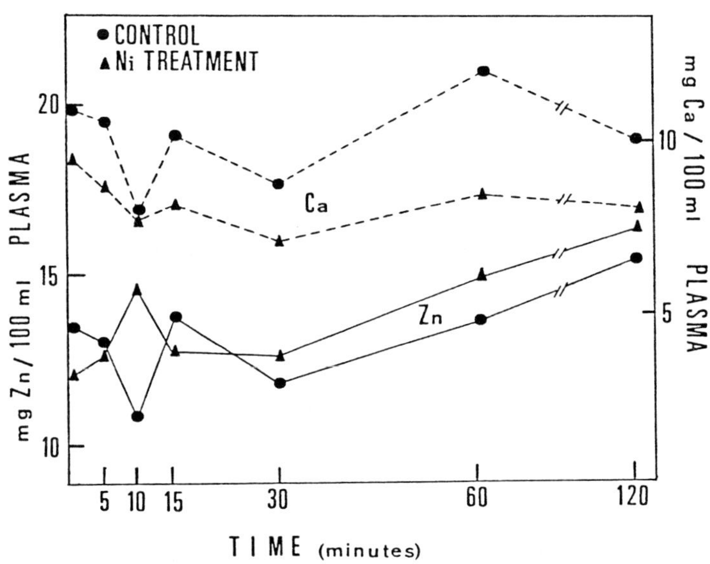

FIG. 1 Zinc and calcium dynamics in control and intraperitoneally nickel injected rats after glucose stimulus.

The animals treated intraperitoneally with nickel showed lower levels of calcemia after glucose injection, which could signify the calcium incorporation by the pancreas. With nickel treatment,the pancreas incorporated, more calcium than in the control animals.

Zinc, like calcium, is affected by the glucose stimulus. The control animals present a decrease of plasma zinc concentration, corresponding to the insulin secretion and subsequent synthesis, increasing afterwards to normal values. Nickel-treated animals after glucose stimulus show an increase of blood zinc concentration, which could express an amplified zinc-insulin response, and repletion of zinc in the pancreas with nickel ingestion. The exact biochemical mechanism of the influence of nickel over and above the other bivalent cations on the metabolism of carbohydrates is not yet clear.

ACKNOWLEDGEMENTS

This work was possible thanks to the help in 1984 of the Consejo Superior de Investigaciones Científicas, Spain, and due to the co-operation of CSIC and CNPq, Brasil.

REFERENCES

Anke M., Grünn M., Dietrich G., Groppel B. and Henning A.(1974) Trace Elements Metabolism in Animals, vol.1, pp.715-718 in Hoekstra W.G., Suttie J.W., Ganther H.E. and Mertz W. (Eds). University Park Press,Baltimore.

Cruikshank D.P.,Pitkin R.M., Donnelly E. and Reynolds W.A. (1981) Obstetric and Gynecology, 58, 430-434.

Cruikshank D.P.,Pitkin R.M. and Reynolds W.A.(1979) American Journal of Obstetrics and Gynecology, 134, 243-247.

Kirchgessner M. and Schnegg A. (1980) Proceedings of Nickel-symposium, Friedrich Schiller Universität Jena, German Democratic Republic.

Rayssiguier Y.,Gueux E. and Weiser D.(1981) Journal of Nutrition 111, 1876-1883.

Ribas Ozonas B., Sánchez Reus M.I.,Rodrigues Lobato N.,García Martín M.C. and Tamarit Torres J. (1981) Proceedings Mineral Elements 80 Part II, pp. 459-464. The Academy of Finland, Helsinki.

Rodrigues N.L., Novelli E.L.B. and Ribas B.(1983) Acta Biológica Leopoldensia,5 43-50.

Sánchez Reus M.I.,García Martin M.C.,Rodrigues N.L. and Ribas Ozonas (1981) Ciencia Biologica, 6,275-278

Sunderman F.W.,Jr.(1977) Annals of Clinical and Laboratory Science,7, 377-398.

CRITICAL REVIEW OF THE EVIDENCE FOR NICKEL(III) IN ANIMALS AND MAN

JACQUELINE E. CROSS, DINAH M. HUGHES and DAVID R. WILLIAMS

Department of Applied Chemistry, UWIST, PO Box 13, Cardiff CF1 3XF, UK

INTRODUCTION

There has been increasing interest during the last decade in the essentiality and toxicology of nickel in vivo (Brown and Sunderman, 1980). At present, little is known concerning the chemical state of nickel in biological species. The metal appears to be in the +II oxidation state in the jack bean enzyme (Dixon et al., 1975) but, in contrast, the nickel centres recently discovered in some bacteria (Lancaster, 1980; Albracht et al., 1982; Teixeira et al., 1983) have been ascribed to the +III oxidation state. These discoveries could herald a surge of research interest into Ni(III) biochemistry. Quantitative analyses of trace metals in biological systems tends to be complicated, and this is particularly so for Ni(III). Thus, before considerable time and money is spent to support Ni(III) researches, it is essential to establish whether Ni(III) is relevant to man's biochemistry.

NICKEL(III) *IN VITRO*

In aqueous aerobic conditions, Ni(III) is immediately reduced to Ni(II) and measurement of the redox potential for the Ni(III)/Ni(II) couple in vitro is exceedingly difficult. Burgess (1976) estimated the redox potential in aqueous solution at +4.2V by comparison with other transition metals.

Iron and cobalt have redox potentials for M(III)/M(II) couple which are considerably lower than that of nickel (+0.77 and +1.82 volts, respectively). However, aqueous ferric and cobaltic species are undetectable in biological systems although thermodynamic considerations suggest that they ought to be present at approximately 10^{-15} - 10^{-20} mol dm^{-3} (May and Williams (1980). Nevertheless, it has been firmly established that both iron and cobalt can exist in the +III oxidation states in vivo when powerfully complexed to polydentate donor groups. Ni(III), having a higher redox potential, might be expected to do so less enthusiastically.

Recent interest in Ni(III) has led to the in vitro preparation of some relatively stable Ni(III) complexes which have been used to predict its likely importance in vivo.

Nickel can be stabilised in the +III state by powerful donor groups. Lati et al. (1972,1975) identified Ni(III) complexes with ethylenediamine, glycine and EDTA which were observed as unstable intermediates in the reactions of hydroxyl radicals with the corresponding Ni(II) complexes. The Margerum research group has investigated a range of peptide complexes of Ni(III) (Bossu and Margerum, 1977) and a comparable set of Cu(III) complexes (Bossu et al., 1977). In general, they conclude that the presence of the peptide considerably reduces the redox potential of the M(III)/M(II) couple to 0.79V-0.96V for nickel and 0.45V-1.02V for copper complexes. In similar studies Chakravorty (1984) prepared and investigated oxime complexes of Ni(III). Most recently, Nieboer (private communication) postulated that human serum albumin (HSA) would stabilise Ni(III). A relatively stable Ni(III)-HSA complex has been prepared by the oxidation of Ni(II) complex with hydrogen peroxide and further studies are presently being done to assess its redox properties.

The stability of Ni(III) complexes in non-aqueous solution and hence, their behaviour in a hydrophobic protein environment, was investigated by Youngblood and Margerum (1980). They measured the redox potentials of M(III)/M(II) couples for the nickel and copper complexes with a deprotonated triglycinamide ($H_{-3}G_3a$) in a range of solvents. Their results show that the redox potential for the Cu(III)/Cu(II) couple decreased whilst those for the Ni(III)/Ni(II) couple increased as the proportion of water was lowered by the addition of a non-aqueous solvent. This suggests that formation of Cu(III) complexes could be enhanced by a protein environment whilst Ni(III) complex formation appears less likely than in an aqueous medium.

Youngblood and Margerum (1980) also considered the stereochemistry of both Ni(III) and Cu(III) complexes and found the latter are more structurally stable. They conclude that since Cu(III) has not been reported in biological systems, the existence of Ni(III) *in vivo* is highly unlikely.

The *in vitro* evidence above indicates that given the right conditions, nickel(III) complexes with biological ligands can be prepared and can remain stable in the absence of oxygen. Such an environment appears to be present in anaerobic bacteria where there is recent evidence to suggest nickel(III) centres have been found which is discussed below.

NICKEL(III) IN BACTERIA

Nickel(III) has been identified in the hydrogenases from three types of bacteria - methanogenic (Lancaster, 1980; Albracht et al., 1982), sulphate reducing (LeGall et al.,

1982; Cammack et al., 1982) and most recently from a photosynthetic bacteria, Chromatium vinosum (Albracht et al., 1983). Lancaster (1980) first proposed Ni(III) after observing the EPR spectra for chemically prepared samples of M.bryantii. He reported the presence of at least five paramagnetic species, which included the expected signals characteristic of iron-sulphur centres. In addition, an unexpected rhombic signal was found, which was originally assigned to Ni(III) by eliminating all other possible species. It was later proven to originate from nickel by using bacteria samples grown on ^{61}Ni-enriched culture, which gave rise to hyperfine structure on the EPR spectra (Lancaster, 1982).

The same characteristic Ni(III) EPR signal has also been observed in hydrogenase from M.thermoautotrophicum (Albracht et al., 1982), Desulfovibrio gigas (LeGall et al., 1982); Cammack et al., 1982; Teixeira et al., 1983) and Chromatium vinosum (Albracht et al., 1983). Cammack et al. (1982) suggested that the Ni(III) centre may be involved in the reaction cycle of the hydrogenase and LeGall et al. (1982) similarly proposed Ni(III) as a catalytic component of the hydrogenase. Both groups have investigated the redox properties of the nickel centre experimentally; however, their results are inconclusive.

It is important to note that the only experimental technique used to prove the existence of Ni(III) *in vivo* has been low temperature (below 80K) EPR spectroscopy. (Cammack et al. (1982) showed that the Ni(III) signal was actually detectable up to 220K.) It has also been suggested that the Ni(III) signal observed might not actually be due to monomeric Ni(III) but rather to a mixed valence dimer Ni(I)-Ni(II) (Thompson, private communication). This would not, however, agree with the one nickel ion per unit of hydrogenase reported by Cammack et al., (1982).

NICKEL(III) IN MAN AND ANIMALS

The +II oxidation state of nickel undoubtedly occurs in man both as a beneficial species and, in some persons, as an industrial contaminant species; however, there have been no claims or evidence produced for nickel(III) occurring in mammals (Brown and Sunderman, 1980). Under normal conditions, neither the aquated Ni(III) ion nor Ni(III) complex will be present and if they were taken into such mammals, perhaps by ingesting Ni(III)- containing bacteria, the degradation of such proteins in the aqueous, oxygen containing environment would soon reduce the Ni(III) to Ni(II).

CONCLUSIONS

A certain category of reducing bacteria appears to contain nickel(III) but there has been no evidence to suggest nickel-(III) as a biologically important state in aerobic conditions. The implications are that the role of this oxidation state _in vivo_ ought not to be overstressed, particularly not at the expense of important nickel researches such as the rôle of nickel(II) in mammals, efforts to improve instrumental techniques to analyze nickel in our diet and in our environment, and those directed towards the detoxification of workers unintentionally contaminated with nickel compounds.

REFERENCES

Albracht,S.P.J., Graf,E.G. and Thauer,R.K. (1982) FEBS Letters, _140_, 311-313.
Albracht,S.P.J., Kalkman,M.L. and Slater,E.C. (1983) Biochimica et Biophysica Acta, _724_, 309-316.
Bossu,F.P. and Margerum,D.W. (1977) Inorganic Chemistry, _16_, 1210-1214.
Bossu,F.P., Chellappa,K.L. and Margerum,D.W. (1977) Journal of the American Chemical Society, _99_, 2195-2203.
Brown,S.S. and Sunderman,F.W.Jr. (1980) (Eds.), Nickel Toxicology, Academic Press, London.
Burgess,J. (1976) Metal Ions in Solution, p.236, Ellis Horwood Ltd., Chichester.
Cammack,R., Patil,D., Aguirre,R. and Hatchikian,E.C. (1982) FEBS Letters, _142_, 289-292.
Chakravorty,A. (1984) Proceedings of International Conference on Coordination Chemistry, Boulder, Colorado.
Dixon,N.E., Gazzola,C., Blakely,R.L. and Zerner,B. (1975) Journal of the American Chemical Society, _97_, 4131-4133.
Lancaster,J.R.Jr. (1980) FEBS Letters, _115_, 285-288.
Lancaster,J.R.Jr. (1982) Science, _216_, 1324-1325.
Lati,J. and Meyerstein,D. (1972) Inorganic Chemistry, _11_, 2397-2401.
Lati,J., Koresh,J. and Meyerstein,D. (1975) Chemical Physics Letters, _33_, 286-288.
LeGall,J., Ljungdahl,P.O., Moura,I., Peck,H.D.Jr., et al. (1982) Biochemical and Biophysical Research Communications, _106_, 610-616.
May,P.M. and Williams,D.R. (1980) in Jacobs,A. and Worwood,M. (Eds.), Iron in Biochemistry and Medicine, Vol.II, pp.1-28, Academic Press, London.
Teixeira,M., Moura,I., Xavier,A.V., DerVartanian,D.V., et al. (1983) European Journal of Biochemistry, _130_, 481-484.
Youngblood,M.P. and Margerum,D.W. (1980) Inorganic Chemistry, _19_, 3068-3072.

LEFT-HANDED STRUCTURES OF ALTERNATING PURINE-PYRIMIDINE POLYNUCLEOTIDES INDUCED BY Ni^{2+} IONS : AN I.R. SPECTROSCOPY STUDY.

PIERRE BOURTAYRE, SAFIA ADAM, JEAN LIQUIER,
JEAN A. TABOURY and ELIANE TAILLANDIER.
Laboratoire de Spectroscopie Biomoléculaire,
U.E.R. de Médecine, 74, rue Marcel Cachin,
93012 Bobigny FRANCE

INTRODUCTION

One of the most striking examples of DNA conformational polymorphism must certainly be the inversion of double helical handedness observed during the B(right) → Z(left) transition of some alternating purine-pyrimidine synthetic oligonucleotides. Immunological techniques have proved the existence of left-handed structures in native systems (for review, see Robert-Nicoud et al., 1984). In dilute solutions, Van de Sande et al. (1982) have shown that nickel chloride induces the B → Z transition of poly(dG-dC). We have observed that Ni^{2+} ions originatingfrom various mutagenic and carcinogenic nickel compounds also stabilize the left-handed configuration of dilute poly(dG-dC) solutions (Bourtayre et al., 1983 ; Liquier et al., 1984). We present here results obtained on the interaction between Ni^{2+} ions and polynucleotides in highly concentrated samples (hydrated films) — conditions which are supposed to be closer to those existing in vivo.

Z FORM OF POLY(dG-dC) IN THE PRESENCE OF Ni^{2+} IONS

The secondary structure of DNAs can be precisely characterized by I.R. spectroscopy, using conformation sensitive absorptions (A, B or Z forms) (for review see Taillandier et al., 1984a, 1984b). The infrared spectrum of poly(dG-dC)-Ni^{2+}, (1 Ni^{2+} per nucleotide), is shown in Fig. 1 (bottom). This spectrum is identical to that obtained with Na^+ counterions in conditions of low hydration and is correlated with the Z helix of the polynucleotide. We can therefore conclude that in condensed phase, as well as in dilute solutions, small amounts of Ni^{2+} ions stabilize the Z conformation of poly(dG-dC).

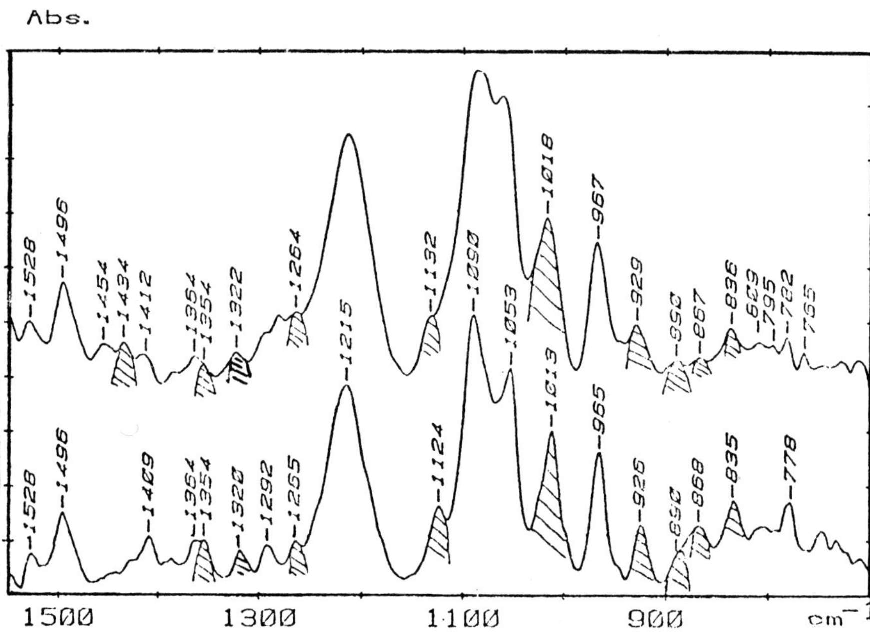

FIG. 1. Infrared absorption spectra of left handed conformations of : top : poly(dA-dC).poly(dG-dT)-Ni^{2+}. bottom : poly(dG-dC)-Ni^{2+}. Absorptions characteristic of the Z form are hatched.

LOCALIZATION OF THE Ni^{2+} INTERACTION SITE

The use of selectively deuterated poly(dG-dC) on the C8 site of guanines allows the investigation of the possibility of an interaction with this base. The I.R. spectrum of poly(dD8G-dC) presents an absorption band located at 1465 cm^{-1} (Figure 2, top) which involves the motion of the N7 nearby atom of the guanine base. This absorption is not conformation sensitive and its modifications will reflect an interaction at the level of the N7 site. In the complex prepared with Ni^{2+} we observe an upward shift of this band to 1476 cm^{-1} (Figure 2 bottom) which shows that among the possible interaction sites on the DNA, the N7 of guanines is involved. The perturbation due to the transition metal ion seems to modify the electronic distribution around the N7 site of the guanine and appears to stabilize the syn conformation of this base. This explains the greater efficiency of Ni^{2+} (and of Co^{2+}), when compared for example to Mg^{2+}, to induce in solution the B → Z transition (Taboury et al., 1984).

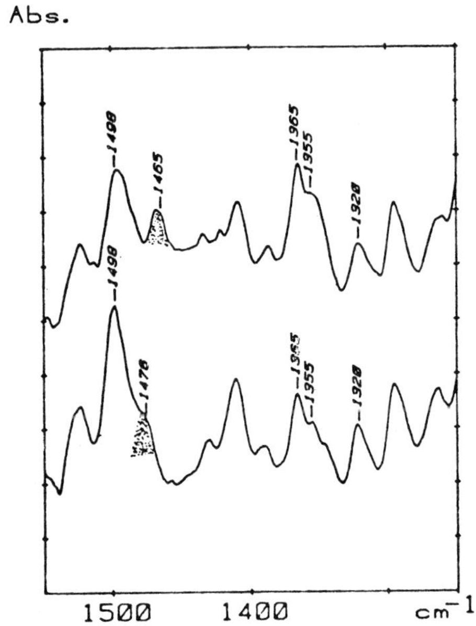

FIG. 2. Infrared absorption spectra of poly(dD8G-dC). Top : sodium counterion ; bottom : Ni^{2+} counterion. The shift of the hatched absorption band proves a strong interaction of Ni^{2+} with the N7 site of guanines.

EFFECT OF Ni^{2+} ON OTHER ALTERNATING PURINE-PYRIMIDINE POLYNUCLEOTIDES.

Long G-C sequences are seldom encountered in native DNAs. On the contrary the ACGT sequence is rather common (Hamada and Hakunaga, 1982), and as it has been suggested that any alternating purine-pyrimidine sequence is potentially able to adopt a left handed structure; we have investigated the effect of Ni^{2+} on the secondary structure of poly(dA-dC). poly(dG-dT) (Taillandier et al., 1984c). Figure 1 presents the I.R. spectra of poly(dG-dC)Ni^{2+} (bottom), and therefore in the Z conformation, and of poly(dA-dC).poly(dG-dT)-Ni^{2+} (top). The absorptions characteristic of the Z helix of poly(dG-dC) (1354 cm^{-1}, 1320 cm^{-1}, 1265 cm^{-1}, 1124 cm^{-1}, 1013 cm^{-1} and 926 cm^{-1}) are found in the poly(dA-dC).poly(dG-dT)-Ni^{2+} spectrum (hatched bands), and we can therefore conclude that in the presence of Ni^{2+} the latter polymer adopts a left handed structure. The I.R. spectrum thus obtained is different from the right handed spectra of poly(dA-dC).poly(dG-dT) (A and B forms) recorded in the presence of Na^+ counterions (figure not shown).

The presence of A-T base pairs in a G-C sequence makes the B → Z transition more difficult to induce in dilute solutions. In the case of poly(dA-dT) we observed the characteristic inversion of the CD spectrum only in strong salt conditions (11 M LiCl, 40 mM $NiCl_2$). In concentrated phase the I.R. spectrum of poly(dA-dT)-Ni^{2+} presents general features consistent with a left handed Z type structure (results to be published).

CONCLUDING REMARK

In conclusion Ni^{2+} ions are able to induce a left handed conformation in regularly arranged alternating purine-pyrimidine sequences : poly(dG-dC), poly(dA-dC).poly(dG-dT) and poly(dA-dT). Among the different mechanisms proposed for the Ni^{2+} carcinogenicity or mutagenicity (for review see Sunderman, 1984), the induction of left-handed structures in native DNA cannot therefore be excluded.

REFERENCES

Bourtayre P., Pizzorni L., Liquier J., Taboury J.A., Labarre J.F. and Taillandier E. (1983) In Proceedings: Nickel in the human environment Lyon 1983, vol. 53. International Agency for Research on Cancer (in press).

Liquier J., Bourtayre P., Pizzorni L., Sournies F., Labarre J.F. and Taillandier E. (1984) Anticancer Research, 4, 41-44.

Robert-Nicoud M., Arndt-Jovin D.J., Zarling D.A. and Jovin T.M. (1984) The EMBO Journal, 3, 721-731.

Sunderman F.W. Jr (1984) in Annals of Clinical and Laboratory Science (in press).

Taboury J.A., Bourtayre P., Liquier J. and Taillandier E. (1984) Nucleic Acids Research, 12, 4247-4258.

Taillandier E., Liquier J., Taboury J.A. and Ghomi M. (1984a) In Spectroscopy of Biological Molecules Sandorfy C. and Theophanides T. (Eds), D. Reidel publ. Co., Dordrecht Holland (in press).

Taillandier E., Liquier J. and Taboury J.A. (1984b) In Advances in Infrared and Raman Spectroscopy Clark R.J.M. and Hester R.E. (Eds), vol. 12. Heyden Publ. Co. London (in press)

Taillandier E., Taboury J.A., Adam S. and Liquier J. (1984c) Biochemistry, (in press).

Van de Sande J.M., McIntosh L.P. and Jovin T.M. (1982) The EMBO Journal 1, 777-782.

ACTIVATION OF THE CALMODULIN-DEPENDENT PHOSPHOPROTEIN PHOSPHATASE BY NICKEL IONS

MARITA M. KING, KIRK K. LYNN, AND CHARLES Y. HUANG

Laboratory of Biochemistry,
National Heart, Lung and Blood Institute,
National Institutes of Health,
Bethesda, MD 20205 USA

INTRODUCTION

The calmodulin-dependent phosphoprotein phosphatase, originally discovered in bovine brain as a calmodulin-binding protein (Wang and Desai, 1977), is now known to catalyze the dephosphorylation of phosphatase inhibitor I, histones, casein, phosphorylase kinase, myosin light chain (Stewart et al., 1982; Yang et al., 1982), the regulatory subunit of cAMP-dependent protein kinase, several brain-specific phosphoproteins (King et al., 1984) and synthetic substrates such as p-nitrophenyl phosphate (PNPP). Although its precise metabolic role is not yet identified, it undoubtedly serves an important function since it participates in reactions involving regulations by phosphorylation-dephosphorylation and by Ca^{2+}-calmodulin. Furthermore, it is the predominant calmodulin-binding protein in the brain. The enzyme consists of two polypeptide chains, a 60,000 M.W. A subunit and a 19,000 M.W. B subunit having four Ca^{2+}-binding sites. It contains one equivalent each of nondissociable Zn^{2+} and Fe^{3+} and can exist in a phosphorylated form (King and Huang, 1984). In addition, it is activated by divalent metal ions such as Ni^{2+}, Mn^{2+}, Co^{2+}, Mg^{2+}, etc. Of these ions, Ni^{2+} exerts the greatest activating effect (King and Huang, 1983; Pallen and Wang, 1984). The present report deals with the mode of activation of the phosphatase by Ni^{2+} ions.

ACTIVATION BY Ni^{2+} USING HISTONE H1 AS SUBSTRATE

When assayed at 30°C, pH 7.6, in the presence of a saturating level of calmodulin (3 μM) and 30 μM [^{32}P]histone H1, Ni^{2+} (300 μM) activated the phosphatase more than 8-fold over the Ca^{2+}-supported activity. Similar activation has been demonstrated with other phosphoproteins, albeit the extent of activation varied with the substrate used. In the absence of calmodulin, the enzyme had little or no activity when only

Ca^{2+} was added, but it displayed high activity upon addition of Ni^{2+}. The activating effect of Ni^{2+} is, therefore, not dependent on calmodulin. Kinetic studies showed that Ni^{2+} primarily affected the maximal rate of dephosphorylation. In the presence of calmodulin, the K_m's for histone H1 obtained with Ca^{2+} and Ni^{2+} were 2.2 and 4.2 μM, whereas the k_{cat}'s were 0.50 and 24 min^{-1}, respectively.

REACTIVATION OF THE CALMODULIN-DEACTIVATED PHOSPHATASE BY Ni^{2+}

Concomitant with activation, Ca^{2+} and calmodulin caused a time-dependent deactivation of the phosphatase, which was accelerated by substrates such as histone H1 and PNPP and by competitive inhibitors such as ADP and PP_i (King and Huang, 1984). The deactivation required the simultaneous presence of Ca^{2+} and calmodulin. Examination of the mechanism of deactivation revealed that it was not due to loss of Zn^{2+} or Fe^{3+}, proteolysis, severe product inhibition, self-dephosphorylation, or dissociation of calmodulin. It was a first-order process consistent with a conformational rearrangement. The deactivation, however, could be reversed, as well as prevented, by divalent metal ions such as Ni^{2+}, Mn^{2+} and Mg^{2+}. These observations indicate that the phosphatase is not simply a Ca^{2+} and calmodulin-stimulated enzyme. In addition to the nondissociable Zn^{2+} and Fe^{3+} and the Ca^{2+} bound to the B subunit and calmodulin, the enzyme requires a divalent metal ion for structural stability and full activity.

MODE OF Ni^{2+} ACTIVATION

Using PNPP as the substrate in a continuous assay, we were able to study the mode of Ni^{2+} activation in greater detail. The main features of Ni^{2+} activation are as follows:

Time-dependence reflecting a conformational change

Activation of the phosphatase by Ni^{2+} was time-dependent, characterized by an initial lag phase (Fig. 1) . The lag phase followed first-order reaction kinetics for more than 3 half-lives, and the lag time (equivalent to the inverse of the first-order rate constant) decreased with increasing Ni^{2+} concentration. A conformational rearrangement of the enzyme following a loose Ni^{2+} binding step is implicated.

Extremely tight binding of Ni^{2+} to the enzyme

The high activity of the Ni^{2+}-activated phosphatase cannot be reduced by chelators like EDTA and EGTA. That the effect of Ni^{2+} is on the enzyme rather than on the substrate is

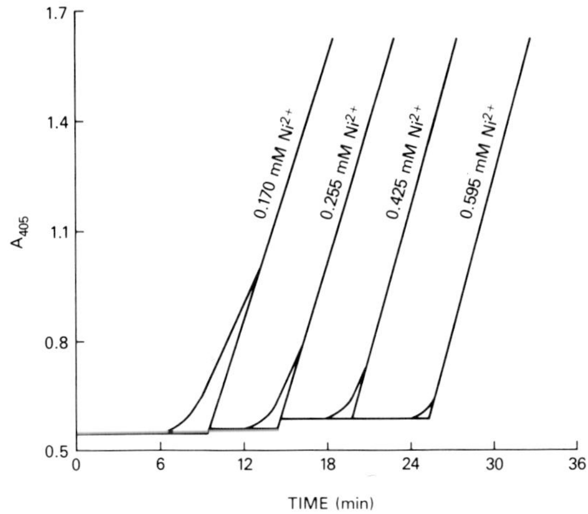

Fig. 1. Activation of calmodulin-dependent phosphoprotein phosphatase as a function of Ni^{2+} concentration. Assay conditions: phosphatase, 0.086 μM; calmodulin, 1.86 μM; PNPP, 10 mM; Ca^{2+}, 0.2 mM; in 50 mM HEPES buffer containing 100 mM KCl, pH 7.6; 30°C. Ni^{2+} (concentration varied as indicated on the progress curves) was added at 6, 12, 18, and 24 min to initiate the activation. Comparable final reaction rates are indicated by the parallel lines. The lag times, defined as the distance between the point where Ni^{2+} was added and the linear line extrapolated from the final phase of reaction, are 3.3, 2.3, 1.7, and 1.3 min as Ni^{2+} increases from 0.170 to 0.595 mM.

evidenced by a 27% quenching of the Trp-fluorescence of the enzyme on adding Ni^{2+}. The quenching also cannot be reversed by addition of chelators. The primary Ni^{2+} binding site is on the A subunit since the dissociated A (prepared by separation on Ultrogel in 8 M urea followed by dialysis) displays similar Trp-fluoresence quenching. The virtually irreversible binding of Ni^{2+} is further supported by the observation that different Ni^{2+} concentrations gave rise to the same final extent of activation, though the lag times of activation were different (Fig. 1).

Acceleration by the presence of substrate

The lag time of Ni^{2+} activation decreased with increasing PNPP concentration, indicating that the enzyme-substrate complex can undergo a more rapid conformational change. This notion is strengthened by the previously observed accelerated deactivation of the enzyme by substrates and competitive inhibitors.

Proposed model

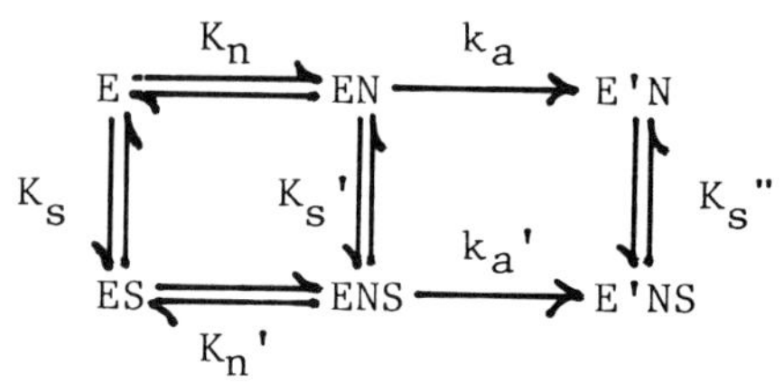

E = enzyme-calmodulin complex

N = Ni^{2+}

S = substrate

The model illustrates initial loose binding steps for Ni^{2+} (K_n and K_n' = 800–2000 μM), followed by irreversible conversion of E to the activated E' forms. $k_a' > k_a$ since the presence of S speeds up the activation process. k_a' is on the order of 2 min^{-1}.

CONCLUSIONS

Ni^{2+} dramatically activates the calmodulin-dependent phosphoprotein phosphatase. The activation process is time-dependent and irreversible. Divalent metal ions like Ni^{2+}, Mn^{2+}, and Mg^{2+} are needed for maintaining the phosphatase's structural stability. If Ni^{2+} plays a physiological role, there may exist a carrier protein to transfer the ion or to serve as a regulator. The irreversible activation of the phosphatase by Ni^{2+} in the presence and absence of calmodulin also suggests that the effect may be toxic by way of interfering with the normal cellular regulation.

REFERENCES

King, M.M., and Huang, C.Y. (1983) Biochem. Biophys. Res. Commun., 114, 955-961.

King, M.M., and Huang, C.Y. (1984) J. Biol. Chem., 259, 8847-8856.

King, M.M., Huang, C.Y., Chock, P.B., Nairn, A.C., Hemmings, H.C., Jr., Chan, K.-F.J., and Greengard, P. (1984) J. Biol. chem. 259, 8080-8083.

Pallen, C.J., and Wang, J.H. (1984) J. Biol. Chem. 259, 6134-6141.

Stewart, A.A., Ingebritsen, T.S., Manalan, A., Klee, C.B., and Cohen, P. (1982) FEBS Lett. 137, 80-84.

Wang, J.H., and Desai, R. (1977) J. Biol. Chem. 252, 4175-4184.

Yang, S.-D., Tallant, E.A., and Cheung, W.Y. (1982) Biochem. Biophys. Res. Commun., 106, 1419-1425.

SUMMARY REMARKS OF SESSION CHAIRMEN (Tor Norseth, Max Costa, Evert Nieboer, Kazimierz Kasprzak, Antero Aitio and David Tonks)

Biological availability, and therefore tissue uptake, seems to be an important factor for risk evaluation of different nickel compounds and this was the theme of the early papers in these sessions. Thus Dr. Hildebrand demonstrated different effects of the highly carcinogenic α-nickel subsulfide and the supposedly less carcinogenic β-nickel sulfide in human embryonic pulmonary cells in culture. Nickel-treated cells were found to be retracted, globular, and with a large number of microvillosities. Nuclei were irregularly shaped and chromatin strands more condensed than in untreated cells. There were, however, important differences between the two compounds in respect of the incorporation of nickel particles - an observation which supports a metabolisation hypothesis for nickel compounds in the development of cancer. However, the possibility remains that differences in particle size after intracellular uptake may account for differences between the two sulfides; the dissolution half-times are different, so that nickel subsulfide is more likely to yield smaller particles. Thus Dr. Hansen reported differences in dissolution rates for each of seven nickel compounds at different pH values. She found that the toxicities of the seven compounds in a cell system were in accordance with the solubility at pH 6 - 7.5 and that they were ranked as follows: black nickel oxides, metallic nickel and ferronickel compounds with green nickel oxides as the least toxic of the compounds tests. These results suggest that the biological effects of particulate nickel compounds may be related to phagocytosis and intracellular dissolution of the particles at the relatively acidic pH of intracellular vacuoles.

Dr. Nieboer discussed the biological reactivity of so-called insoluble nickel compounds. Surface activity of the particles was tested by hemolytic ability, and comparisons of nickel particles of various carcinogenic potentials indicated that the surface passivity of relatively insoluble nickel compounds might be an important determinant in nickel carcinogenesis.

Dr. Stern described a device specifically designed to collect welding aerosols for use *in vitro* genotoxicity assays. It is well known that particles of nickel and chromium may undergo chemical or physical alterations before and during collections for biological assays, but with the system described such alterations were minimized. Pilot studies demonstrated that no new toxic species were generated in the system, whereas active nickel and chromium species may behave differently when collected on a membrane filter or by other traditional methods based upon liquid systems. *In vivo* and *in vitro* studies of the comparative toxicities of nickel, chromium and other metal salts to the lung were described by Dr Benson.

Dr Kodama reported on the pulmonary deposition and clearance of inhaled nickel oxide. A dose-response relationship was found between the estimated inhaled nickel dose and the measured lung burden of nickel oxide. As expected, the deposition fractions varied with increasing aerodynamic diameter but the deposition fractions were not affected significantly by exposure concentration or total exposure time. Nickel was apparently not absorbed from the lungs, as there were no increases in the nickel burdens of liver, kidney, spleen, heart, brain or blood. Experiments designed to uncover the chronic effects of inhaled nickel oxide were described by Dr Takenaka, who showed that rats developed severe alveolar proteinosis.

The discussion of these findings revealed a number of difficulties with the reported experimental models and some disagreements between present and previously reported results, which are probably related to methodological differences. However, such studies are probably the only way to get further information about the relative potency of different nickel compounds. Epidemiological investigations and animal experiments are weak tools for such investigations: only high risk levels can be investigated in animal experiments, and specific exposure to single compounds is very rare in the nickel refining industry.

The papers which described the actual or potential biochemical interactions of nickel compounds ranged widely in their scope, and implicated lipid peroxidation (Prof Sunderman), erythropoietin (Dr Hopfer), glucose metabolism (Dr Ribas), and binding to DNA (Dr Bourtayre) and bacterial chromatin (Dr Sigee). Novel findings were presented by Dr Huang on the activation of the calmodulin-dependent-phosphoprotein, calcineurin, by various divalent cations, particularly Ni^{2+}. Activation by Ni^{2+} was observed in the presence and absence of calmodulin and it was not reversed by high concentrations of chelators. These observations indicate that Ni^{2+} ions bind very tightly to calcineurin. The results show that calcineurin is not simply a Ca^{2+} and calmodulin-stimulated enzyme: the enzyme requires a divalent metal ion for structural stability and full activity.

CLINICAL PATHOLOGY

461634

SIMULTANEOUS DETERMINATION OF NICKEL AND CADMIUM IN WHOLE BLOOD BY ZEEMAN ATOMIC ABSORPTION SPECTROPHOTOMETRY: BLOOD NICKEL AND CADMIUM LEVELS IN THE GENERAL POPULATION OF DENMARK

J. MOLIN CHRISTENSEN* and M. KIRCHHOFF**

*Danish National Institute of Occupational Health, DK 2900 Hellerup, Denmark

**Dep. of Internal Medicine, C Glostrup Hospital, University of Copenhagen, DK-2600 Glostrup, Denmark

INTRODUCTION

Humans are exposed to nickel and cadmium in variable amounts. In the industry nickel and cadmium are used as metals, catalysts, pigments and in Ni/Cd batteries. Thus, for determinations of occupational exposure, metabolism and biochemistry there is much interest in monitoring cadmium and nickel in blood.

Most of the previously published Ni/Cd blood results are too high due to contamination and analytical difficulties (Sunderman et al., 1980; Stoeppler M., 1983). Therefore it is highly desirable to develop a method which gives accurate and rapid quantitation of cadmium and nickel at normal blood levels. Several techniques such as resin adsorption, extraction and acid digestion have been applied. Sunderman et al. (1984) reported a sensitive, direct method for determination of nickel in whole blood using matrix modification, deproteinisation, heating and Zeeman background correction.

The paper presented here describes a direct method with simple dilution for simultaneous determination of cadmium and nickel in whole blood using $Cr(NO_3)_3$ as matrix modifier, Perkin Elmer atomic absorption spectrophotometer and Zeeman effect for background correction.

METHODS AND RESULTS

Apparatus and Instrument Settings.

A Perkin-Elmer Model 5000 atomic absorption spectrophotometer equipped with a ZEEMAN graphite furnace HCA-500 programmer, AS-40 autosampler and pyrolytically coated furnace tubes were used. High speed data were collected by a Perkin Elmer 7300 data station and displayed using a PR-100 printer.

The instrumental parameter settings were: Wavelength 228.8 nm for cadmium and 232.0 nm for nickel; slit "low" 0.7 nm; scale expansion x5. Furnace settings for cadmium: drying at 90 to 250°C in 30 sec; atomization at 2100°C for 6 sec. Furnace settings for nickel: drying at 90 to 250°C in 75 sec., charring at 1200°C for 20 sec., and atomization at 2600°C for 8 sec. Argon gas-stop mode was used.

Materials

Cadmium and nickel standard solutions1.00 g/l (BDH Chemicals, England). Minisorb[R] polyethylene tubes and Cryo polypropylene tubes from Nunc (Denmark).

Sample handling

All glassware, ployethylene tubes and sample cups were cleaned for 1 hour in a 1 M HNO_3 solution and rinsed with deionized water just before use. Heparinized whole blood was obtained by venepuncture using "metal free" Vacutainer[R] tubes and polyethylene cannula. The cannula was flushed with 2-3 ml of blood, which was discarded. To investigate contamination from Vacutainer[R] tubes whole blood was collected with "Monovette" (Sarstedt) syringe and polyethylen intraveneous cannula. No significant differences were observed between nickel concentration in blood collected in vacutainer tubes versus "Monovette" syringe. Heparinized whole blood was stored at -80°C.

Procedure for whole blood

Duplicate 0.5 ml aliquots of blood were transferred to Minisorb[R] tubes containing 150 µl Triton x-100/citrate (2.5%/3.8%). 500 µl 2% $Cr(NO_3)_3$ and 1.5 ml HNO_3 were added; the tubes were vortex mixed for 10 sec and centrifuged at 2000 xg for 15 min. The supernatant was recentrifuged. Sixty µl of the supernatant were pipetted into the graphite tube. Ni/Cd blood standards were prepared in the concentration range 1.5, 3.0 and 6.0 µg/l.

A blood standard curve is constructed by linear regression and sample blood concentration calculated from peak height. The sensitivity for cadmium and nickel in blood was 0.1 µg/l and 0.2 µg/1, respectively.

The addition of $Cr(NO_3)_3$ lead to a rapid decomplexation of nickel and cadmium with removal from binding sites and allow nickel and cadmium to be measured without heating the samples (fig. 1).

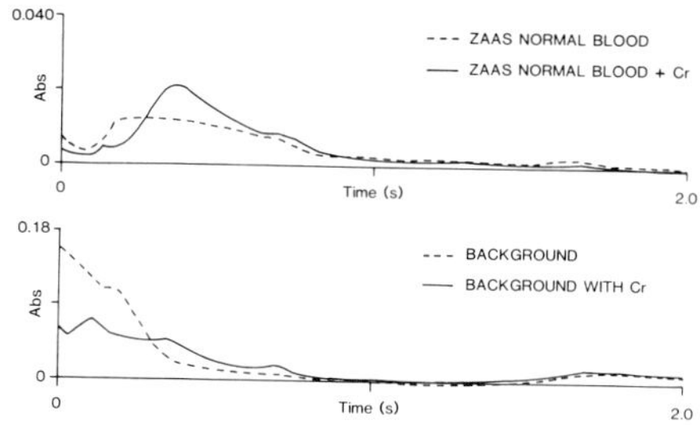

FIG.1. The influence of $Cr(NO_3)_3$ on determination of Nickel in blood.

Furthermore $Cr(NO_3)_3$ has a significant influence on the absorbance profile of nickel (fig. 2).

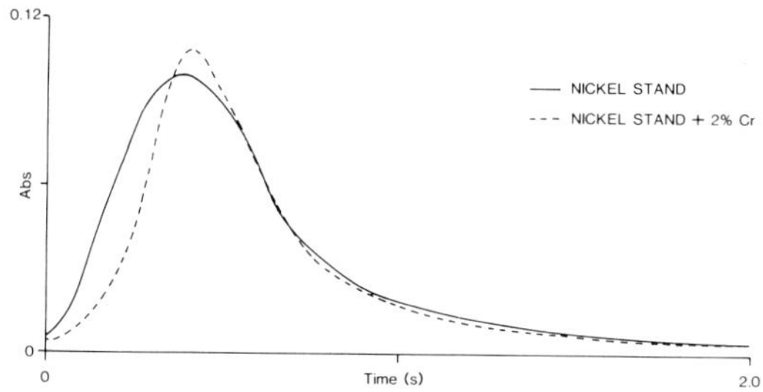

FIG.2. The effect of $Cr(NO_3)_3$ on the absorbance profile of nickel.

Precision.

As control material a pool sample of blood was collected. A high pool was made by spiking the whole blood with aqueous cadmium and nickel standards. The controls were run every day of analysis. At a Cd concentration of 0.33 µg/l and Ni concentration 1.30 µg/l the coefficients of variation for 10 blood samples were 9.1% and 5.3% respectively.

Interferences.

The possible effects of endogenous metals in blood were tested, but no interferences were observed for Fe, Mn, Zn and Cu.

Comparison studies.

We analysed control materials for cadmium and nickel. Control blood was obtained from the Behring Institute: Seronorm[R] trace element and Seronorm animal-blood from Nyegaard, Oslo.

TABLE 1. Cadmium and nickel concentrations in control materials analysed by AAS-ZEEMAN.

Sample	N	Cadmium (µg/l)		Nickel (µg/l)	
		Mean±SD	Asigned Value	Mean±SD	Asigned Value
Behring Blood 1	25	5.6±0.3	5.5±0.7	7.1±0.4	5-10
Seronorm[R] TM	25	0.8±0.1		3.1±0.2	3.3±0.3*
Seronorm[R] animal-blood**	35	0.8±0.2		1.2±0.2	

* Performed at the Dep. of Lab. Med. and Pharm. University of Connecticut School of Med. USA.

** Supplied by Y. Thomassen, Inst. of Occup. Health, Oslo.

Reference values.

The normal values for cadmium and nickel in human blood from 25 nonexposed 40 years old men were between 0.4 - 2.3 µg/l (median 0.7 µg/l), and 0.8-1.9 µg/l (median 1.4 µg/l respectively.

CONCLUSIONS

A new sensitive, precise and rapid procedure for analysis of Cd and Ni in blood simultaneously was developed using EAA/ Zeeman background correction. The accuracy of the method is satisfactory as indicated by our results obtained by analysis of control blood. The simplicity of the procedure results in decreased risk of nickel contamination and makes the method attractive for routine use.

REFERENCES

Ostapczuk P., Valenta P., Stoeppler M. and Nürnberg H.W. (1983) in Brown S.S. and Savory J. (Eds.), Chemical Toxicology and Clinical Chemistry of Metals, pp 61-64. Academic Press, London.

Sunderman F.W., (1980) Pure. & Appl. Chem 52 pp. 527-544.

Sunderman F.W., Christina C., Reid M.C. Hopfer S.M. and Nomoto S. (1984) Ann. Clin. Lab. Sci, in press.

U61635

SQUARE WAVE VOLTAMMETRY: A NEW APPROACH FOR THE SENSITIVE DETERMINATION OF NICKEL AND COBALT IN HUMAN SAMPLES

PETER OSTAPCZUK, MARTINA FRONING, MARKUS STOEPPLER and HANS WOLFGANG NÜRNBERG
Institute of Applied Physical Chemistry,
Nuclear Research Center (KFA) Juelich, P.O. Box 1913,
D-5170 Juelich, Federal Republic of Germany

INTRODUCTION

Square wave polarography at a dropping mercury electrode was described first by Barker and Jenkins (1952). Its application, however, was limited by a lack of appropriate commercial instrumentation. Very recently, instruments have been introduced that allow the routine use of this promising approach. Thus, we determined if adsorption square wave voltammetry (ASWV) could be of advantage for nickel and cobalt determination in digestion solutions of biological samples after interfacial accumulation of an adsorption layer of heavy metal chelates at the hanging mercury drop electrode (Pihlar et al., 1981).

EXPERIMENTAL

Instrumentation

A Polarographic Analyzer System consisting of a model 384 B console, DMP-40 digital plotter, model 303 A SDME and model 305 stirrer, all from EG & G, Princeton, N.J., U.S.A., has been used. The 384 B is a microprocessor-based system, which offers eight different electrochemical techniques and an automated determination.

Sample treatment

The universally applicable wet digestion procedure used for various biological materials involving a mixture of nitric and perchloric acids has been already described elsewhere. It permits the complete decomposition of materials as whole blood, blood serum, urine, nails, hair, tissue and bones (Ostapczuk et al., 1984).

Determination step for nickel and cobalt

5 ml of the blank analyte (0.05 M ammonia buffer with 1×10^{-4} M dimethylglyoxime) are put into the voltammetric cell and deaerated for 4 min with pure argon (99,999 %). After setting an adsorption potential

of -0.7 V versus the Ag/AgCl reference electrode for an adsorption time of about 30 sec, nickel and cobalt dimethylglyoximates are adsorbed at the SDME as a film under stirring at 800 rpm. After a rest period of 5 sec, the background voltammogram is recorded in the square wave mode (pulse height 20 mV, staircase step size 2 mV, frequency 100 Hz) by scanning with 200 mV/s into the negative potential direction. The blank curve is stored automatically. Subsequently up to 1 ml of sample solution is added. After an additional deaeration of 90 sec the whole procedure, as described before for the blank determination is repeated and the voltammogram so obtained is stored. The concentrations of nickel and cobalt determined in this way we automatically evaluated on the basis of two standard additions.

RESULTS AND DISCUSSION

It has been reported already that SWV is a highly sensitive technique for lead and cadmium determinations aqueous solution (Buchanan and Soleta, 1983), but no information has been available until now about the behaviour of nickel and cobalt. With the conditions described above, we observed that nickel can be determined with about forty-fold increased sensitivity (somewhat depending on pre-treatment effectiveness) compared to the differential pulse mode (Ostapczuk et al. 1983) and that this figure for cobalt is

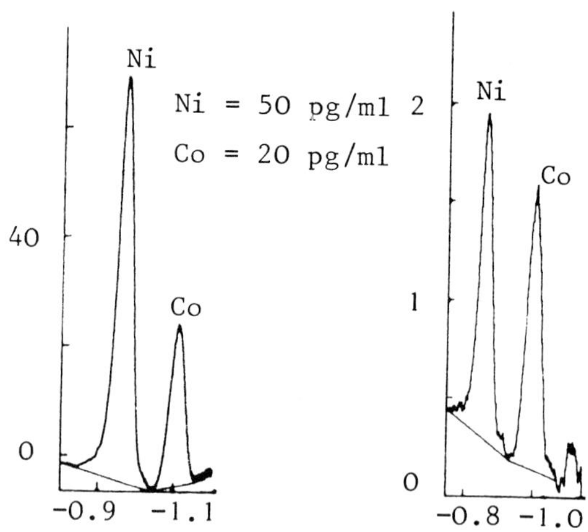

FIG. 1. Ni and Co determination by ASWV (a) and ADPV (b)

about twenty-fold (Fig. 1). An additional advantange is the fast scan rate, reducing the time needed for determination of both elements to about 50 % compared to ADPV (Ostapczuk et al. 1983). Studying the parameters that influence the peak current we found that increasing peak currents of nickel and cobalt increased the pulse height, frequency and staircase step size.

An important advantage due to this enhanced sensitivity is that now analysis is possible even after an incomplete digestion.

TABLE 1. Ni and Co in body fluids and tissue samples from persons with metal implants

Specimen	n	Nickel Mean µg/kg	Nickel Range µg/kg	Cobalt Mean µg/kg	Cobalt Range µg/kg
Whole blood	8	3.66	1.25-11.4	0.28	0.10-0.56
Serum	8	6.19	1.77±23.7	0.55	0.15-1.54
Urine	7	5.47	0.95-19.4	1.43	0.15-8.07
Bone	4	5250	260-17800	25170	380-59500
Tissue	5	14550	260-64000	24200	24-117700
Finger nails	19	1850	360-7820	187	34-1000
Finger nails (unexposed)	5	1380	340-3210	90	20-190
Toe nails	15	390	220-700	78	21-230
Toe nails (unexposed)	5	290	180-390	31	25-39
Hair	28	1450	170-8090	174	31-1120
Hair (unexposed)	7	1250	490-1870	80	59-100

Table 1 presents an overview on recently determined nickel and cobalt levels by application of the described method in various samples from persons with stainless steel implants and unexposed control persons within the current research project in cooperation with INSERM Lille (Hildebrand et al. 1984). The results show that nickel contents are elevated in exposed persons and that this metal mainly is found in blood serum, probably bound to serum proteins as has been observed earlier by Herlant-Peers et al. (1982) in studies with ^{63}Ni. The same was found for cobalt and for both elements in urine, hair, finger and toe nails in comparison to normal levels (Sunderman, 1980; Ostapczuk et al., 1983). From results obtained so far it appears that the most promising indicator media for further studies are whole blood and serum.

The increase observed for cobalt in nails and hair, however, was much more significant than for nickel. Nickel and cobalt concentrations in bone and tissue showed the expected dependence on the position relative to the implants. The comparatively high cobalt levels found for bone and tissue require an additional determination of normal levels for these materials, and this study will be performed in the near future.

ACKNOWLEDGEMENT
This work has been financially supported by Deutsche Forschungsgemeinschaft under contract Sto 149/2-1 which is gratefully acknowledged.

REFERENCES

Barker G.C.& Jenkins I.L (1952) Analyst, (London) 77, 685-696.
Buchanan E.B.& Soleta D.D. (1983) Talanta, 30, 459-464.
Herlant-Peers M.C., Hildebrand H.F.& Biserte G. (1982) Zbl. Bakt. Hyg. I. Abt. Orig. B176, 268-382.
Hildebrand H.F., Roumazeille B., Decoulx J., Herlant-Peers M.C., Ostapczuk P. and Stoeppler M. (1984). These Proceedings, p. 169
Ostapczuk P., Valenta P., Stoeppler M., Nürnberg H.W. (1983) in S.S. Brown and J. Savory (Eds.) Chemical Toxicology and Clinical Chemistry of Metals, pp. 61-64. Academic Press, London.
Ostapczuk P., Goedde M., Stoeppler M.& Nürnberg H.W. (1984) Fresenius Z. Anal. Chem., 317, 252-256.
Sunderman F.W. (1980) Pure Appl. Chem., 52, 527-544.

461636

HYPERNICKELEMIA IN HEMODIALYSIS PATIENTS

SIDNEY M. HOPFER,* JEANNE V. LINDEN,* CRISTINA CRISOSTOMO,* FRANK A. CATALANATTO,** MALCOLM GALEN*** and F. WILLIAM SUNDERMAN Jr.*
*Department of Laboratory Medicine, **Department of Pediatric Dentistry, University of Connecticut Health Center, Farmington, CT 06032, and ***Department of Medicine, St. Francis Hospital, Hartford, CT 06105, USA

INTRODUCTION

Investigators who have surveyed trace metal concentrations in serum or whole blood of hemodialysis patients have generally focused attention on the increased Al concentrations and have neglected to measure Ni concentrations. Salvadeo et al. (1979) noted higher Ni concentrations in fluid entering the dialyzer compared to that leaving the dialyzer, suggesting that patients may experience Ni uptake during hemodialysis. Webster et al. (1980) reported acute Ni intoxication in 23 dialyzed patients owing to dialysis fluid that was contaminated with Ni from a heater tank. The patients' symptoms of nausea, vomiting, weakness, headache, and palpitation disappeared within 13 hours after cessation of dialysis. In the present study, because of our concern about possible iatrogenic nickel poisoning (Sunderman, 1983), Ni concentrations were determined in body fluids of patients with end-stage renal disease treated by hemodialysis.

MATERIALS AND METHODS

This study was conducted at three hemodialysis facilities in central Connecticut (Table 1). Serum and whole blood specimens were obtained from (a) 30 healthy control subjects, age 25 to 84 years, (b) 7 patients with end-stage renal disease who had not been treated by hemodialysis, and (c) 59 patients with end-stage renal disease who were receiving hemodialysis treatments 3 times/week (3-4 hours/treatment). Pre-dialysis specimens of blood were collected from the arterial catheter prior to heparinization; post-dialysis specimens were collected from the arterial catheter at the end of hemodialysis. Specimens of parotid saliva were collected by suction from Stensen's duct (Catalanatto et al., 1977). Ni concentrations in serum, whole blood, and saliva were analyzed directly by electrothermal atomic absorption spectrometry with Zeeman background correction (Sunderman et al. 1984). Ni concentrations in 10 ml samples of water and electrolyte concentrate solutions were analyzed after acid digestion, chelation, and solvent extraction (Brown et al. 1981).

TABLE 1. Description of the hemodialysis facilities

Parameter	Hosp A.	Hosp B.	Hosp C.
No. of stations:	15	6	9
Reservoir for dialysis solution:	central tank	individual tanks	individual tanks
Hemodialysis apparatus:	Drake-Willock Co. model 4509	Cobe Labs. model 018700	Gambro, Inc. model AK-10
Dialysis membrane:	Cuprophan[a]	Cuprophan[a]	Cuprophan[a]

[a]Travenol hollow fiber dialyzer (cuprammonium rayon) (model CF 15-11), blood flow = 200-300 mL/min, dialysis solution flow = 500-550 mL/min, surface area = 1.0 m^2.

TABLE 2. Nickel concentrations in serum and whole blood

Population	No. of subjects	Nickel Conc. (µg/L, mean ± SD) serum	whole blood
Healthy Controls	30	0.28 ± 0.24	0.34 ± 0.28
End-Stage Renal Dis.	7	0.57 ± 0.25	0.67 ± 0.47
Hospital A	40		
Pre-dialysis		6.2 ± 1.8[a]	4.5 ± 1.6[a]
Post-dialysis		7.2 ± 2.2[b]	5.4 ± 1.9[b]
Difference		1.0 ± 1.1 (+16%)	0.9 ± 1.1 (+20%)
Hospital B	9		
Pre-dialysis		3.9 ± 2.0[a]	2.9 ± 1.7[a]
Post-dialysis		5.2 ± 2.5[b]	3.6 ± 1.7[b]
Difference		1.3 ± 1.0 (+33%)	0.7 ± 0.7 (+24%)
Hospital C	10		
Pre-dialysis		3.0 ± 1.3[a]	2.5 ± 1.0[a]
Post-dialysis		3.7 ± 1.3[b]	2.8 ± 0.8[b]
Difference		0.7 ± 0.3 (+23%)	0.3 ± 0.3 (+12%)

[a] $P < 0.001$ vs healthy controls.
[b] $P < 0.01$ vs pre-dialysis value (paired t test).

TABLE 3. Nickel concentrations in pre-dialysis specimens from patients at Hospital A

Months of dialysis	No. of subjects	Ni concentration (µg/L) (mean ± SD) serum	whole blood
1 - 12	10	5.1 ± 1.9[a]	3.7 ± 1.9
13 - 36	7	6.5 ± 1.6	4.8 ± 1.0
37 - 74	11	6.4 ± 1.6	4.8 ± 1.2
75 - 108	7	7.1 ± 1.2	4.9 ± 1.9
109 - 132	7	6.1 ± 2.0	4.8 ± 1.5

[a] $P < 0.05$ vs patients with 13-132 mos. of dialysis therapy.

TABLE 4. Nickel concentrations in hemodialysis solutions

Trade-name of electrolyte concentrate	Ni concentration (µg/L) electrolyte concentrate[a]	water for hemodialysis	hemodialysis solution
Hosp. A			
Eri-Lyte 8129 (Feb 1984)	18.7	0.28	0.82
Eri-Lyte 8129 (May 1984)	18.7	0.04[b]	0.53
Hosp. B			
Eri-Lyte 8135	13.4	0.05	0.47
Eri-Lyte 8118	11.9	0.05	0.40
Hosp. C			
Eri-Lyte 8324	17.6	0.15	0.68

[a] Electrolyte concentrate solutions (Erika, Inc., Rockleigh, NJ) are diluted 34-fold with water.
[b] Hosp. A installed a new reverse-osmosis water system in March 1984.

TABLE 5. Ni concentrations in pre-dialysis specimens of serum and whole blood from the same hemodialysis patients over an interval of 5-6 months

Patients	No.	Body fluid	Ni Concentration (µg/L; mean ± SD) Nov-Dec 1983	May-June 1984	difference
Hosp. A	6	S	6.9 ± 2.3	4.5 ± 1.1[a]	2.3 ± 1.8 (-34%)
		WB	4.4 ± 1.7	3.3 ± 0.7[a]	1.2 ± 1.1 (-27%)
Hosp. B	6	S	5.9 ± 2.2	5.0 ± 1.9	0.8 ± 0.8 (-14%)
		WB	4.3 ± 1.8	3.7 ± 1.9	0.5 ± 1.4 (-12%)

[a] $P < 0.01$ versus prior value by paired t test.

RESULTS

Hypernickelemia (> 1.1 µg/L) was found in all hemodialysis patients. Ni concentrations in serum and whole blood were significantly higher in post-dialysis specimens than in the corresponding pre-dialysis specimens (Table 2). The increase in Ni concentrations was not attributable to hemoconcentration, since serum albumin concentrations and blood hematocrit values were only 8% higher in post-dialysis specimens than in the corresponding pre-dialysis specimens, based upon measurements in 9 patients at Hospital B. Ni concentrations in parotid saliva from 8 hemodialysis patients at Hospital A (0.11 ± 0.11 µg/L) did not differ from values in 12 healthy controls (0.11 ± 0.08 µg/L). After the first 12 months, hypernickelemia was not related to the duration of hemodialysis therapy (Table 3). Three months after a new reverse-osmosis system was installed at Hospital A, Ni concentrations in the hemodialysis solution were reduced 35% and Ni concentrations in patients' serum and whole blood specimens were reduced 34% and 27%, respectively (Tables 4 and 5).

DISCUSSION

This study demonstrates that serum Ni concentrations in hemodialysis patients at three hospitals average 11 to 22 times the mean Ni concentration in serum from healthy subjects. The authors speculate that hypernickelemia may contribute to hemodialysis-associated hypersensitivity, encephalopathy, and arteriosclerosis; they recommend that Ni concentrations in dialysis solutions be reduced to < 0.2 µg/L. To reach this target, Ni concentrations in the 34x electrolyte solution should be < 5 µg/L; Ni concentrations in reverse-osmosis water should be < 0.05 µg/L.

REFERENCES

Brown S.S., Nomoto S., Stoeppler M. and Sunderman F.W. Jr. (1981) Clin. Biochem., 14, 295-299.

Catalanatto F.A., Sunderman F.W. Jr. and MacIntosh T.R. (1977) Ann. Clin. Lab. Sci., 7, 146-151.

Salvadeo A., Minoia C., Segagni S. and Villa G. (1979) Int. J. Artif. Organs, 2, 17-21.

Sunderman F.W. Jr. (1983) Ann. Clin. Lab. Sci. 13, 1-4.

Sunderman F.W. Jr., Crisostomo C., Reid M.C., Hopfer S.M. and Nomoto S. (1984) Ann. Clin. Lab. Sci., 14, 232-241.

Webster J.D., Parker T.F., Alfrey A.C., Smythe W.R., Kubo, H., Neal G. and Hull A.R. (1980) Ann. Intern. Med. 92, 631-633.

SERUM AND LYMPHOCYTE NICKEL AND ALUMINIUM CONCENTRATIONS IN PATIENTS WITH EXTRACORPOREAL HEMODIALYSIS

JOHN SAVORY, SUE BROWN, ROGER L. BERTHOLF, RICHARD ROSS AND MICHAEL R. WILLS.
Depts of Pathology, Biochemistry & Internal Medicine, University of Virginia Medical Center, Charlottesville, Virginia 22908, USA

INTRODUCTION

The accumulation of aluminium is known to occur in patients with chronic renal failure on long-term treatment with hemodialysis. Such accumulation has been proposed to be a major, if not the major, factor in the pathogenesis of a variety of clinical disorders associated with long-term treatment (Wills and Savory, 1983). Most attention has been focussed on aluminium as the major toxicological factor in these patients; however, tissue accumulation of other trace metals may be of significance. Increased serum nickel concentrations occur in patients with chronic renal failure (Sunderman, these proceedings, p145). In view of the known toxic effects of nickel an associated increase in tissue content in patients with chronic renal failure could be of clinical importance.

A major problem in the diagnosis of trace metal poisoning, such as with aluminium or nickel, is that serum concentrations may not provide a valid index of either tissue concentration or total body burden. In the study reported here lymphocyte concentrations of aluminium and nickel were measured as an index of tissue concentration for comparison with simultaneously measured serum values. Measurements were made in normal control subjects and in patients with chronic renal failure on long-term treatment with hemodialysis. Lymphocytes were chosen for this study because they could be isolated as a "tissue" with a population of single cells having a constant nuclear to cytoplasmic ratio in contrast to the mixed nature of leucocytes - a mixed population of cells with a variable ratio.

MATERIAL AND METHOD

Blood for normal control values (serum and lymphocytes) was drawn from healthy adult volunteers using a butterfly needle and a plastic syringe. Blood for lymphocyte separation was dispensed into two 10 ml sodium heparin Vacutainer tubes (no. 6527, Royal Blue top, Becton Dickinson and Co.,

Paramus, NJ) and blood for serum aluminium and nickel concentrations was dispensed into a 17 x 100 mm plastic tube and capped (Falcon, Oxnard, CA). Blood from hemodialysis patients was allowed to flow from tubing connected to the patient's venous fistula directly into the plain and heparin tubes described above. All of the specimens from the dialysis patients were collected immediately before dialysis.

Serum aluminium concentrations were measured by electrothermal atomic absorption (EAAS) using the method of standard additions (Brown et al., 1983). The concentrations of nickel in serum were measured by EAAS (Sunderman et al., 1984) following protein precipitation with nitric acid. Lymphocytes were separated from heparinized whole blood using a density gradient (Histopaque-1077, Sigma Chemical Co., St. Louis, MO) and centrifugation technique. The separation procedure was performed within 2 hours after blood collection. The isolated lymphocyte layer was removed and treated with a lysing reagent (ammonium chloride and potassium bicarbonate) to remove contaminant red blood cells. This mixture was centrifuged at a low rpm, and the resulting pellet (lymphocytes) was washed three times with phosphate-buffered saline (PBS). The concentration of cells in a known amount of PBS was calculated from a cell count performed using a Spencer Bright Line hemocytometer and Turk's counting solution. A Wright's stain of the lymphocyte pellet from some of the controls showed that it contained 95 to 100% lymphocytes. A standard lymphocyte cell concentration for the final analysis was achieved by adding an amount of water to the pellet calculated from the actual cell concentration so as to give the required value (Whitehouse et al., 1982). Water was used as the lysing agent for the lymphocytes, and two freeze/thaw cycles were performed to insure complete lysis (Pleban and McHugh, 1983). The resulting supernate was analyzed for aluminium and nickel by EAAS. Concentration units are reported in $\mu g/10^{10}$ cells (Whitehouse et al., 1982).

RESULTS AND DISCUSSION

The results of the aluminium and nickel measurements in serum and lymphocytes are shown in Table 1 for both the control subjects and the patients on hemodialysis.

TABLE 1. Measurements of Al and Ni

		Controls	Patients
Serum Al	n	50	28
(µg/L)	$\bar{x} \pm$ SD	6.1 ± 2.7	72.3 ± 61.7*
Lymphocyte Al	n	13	27
(µg/10^{10} cells)	$\bar{x} \pm$ SD	0.97 ± 0.47	4.76 ± 2.55*
Serum Ni	n	18	28
(µg/L)	$\bar{x} \pm$ SD	0.44 ± 0.18	3.71 ± 1.54*
Lymphocyte Ni	n	10	22
(µg/10^{10} cells)	$\bar{x} \pm$ SD	0.72 ± 0.75	0.99 ± 1.02^{NS}

* $p < 0.00005$ when compared with controls.

^{NS} Not significantly different.

The correlation data for these same serum and lymphocyte measurements on dialysis patients are shown in Table 2.

TABLE 2. Correlations of Al and Ni measurements

X	Y	n	r
Serum Al	Lymphocyte Al	28	0.353
Serum Al	Serum Ni	28	0.007
Serum Ni	Lymphocyte Ni	23	0.140
Serum Al	Lymphocyte Ni	23	-0.045
Lymphocyte Al	Lymphocyte Ni	23	0.068

Table 1 as expected shows that serum aluminium was significantly elevated in the hemodialysis patients. This observation has been documented in hundreds of hemodialysis centers and results from transfer of aluminium from dialysis solution, oral intake of aluminium especially from phosphate binding gels, and problems with urinary excretion. In all of the patients studied, an increased serum aluminium concentration was associated with an increased lymphocyte content; however, the extent of the elevations was poorly correlated as shown in Table 2. It could be proposed that the lymphocyte values are a better representation of an increase in tissue content and total body burden in these patients, than are serum aluminium concentrations.

Serum nickel was elevated significantly in the hemodialysis patients but surprisingly there was no correlation between these elevations and those for serum aluminium.

CONCLUSIONS

Thus, there appears to be uptake and/or reduced elimination of nickel in hemodialysis patients but presumably the source(s) and mechanisms for removal are different to aluminium. Contamination of dialysis solutions by nickel certainly exists. It is probably not related to aluminium contamination which results mostly from water treatment with aluminium. Also, the dosage of oral intake of nickel will be different to aluminium which is administered in massive amounts to most hemodialysis patients as a phosphate binder. There was no increase in lymphocyte nickel concentrations which might be consistent with the proposal that, despite increased serum concentrations, there is no increase in tissue content in patients with chronic renal failure.

The relative value of serum versus lymphocyte aluminium in assessing body burden remains to be clarified and will probably require tissue analyses. Similarly the question of the extent of nickel loading in hemodialysis patients also will require tissue measurements. Any clinical sequelae resulting from hypernickelemia also remain to be evaluated.

REFERENCES

Brown S, Bertholf R.L., Wills M.R. and Savory J. (1984) Electrothermal atomic absorption spectrometric determination of aluminum in serum using a new protein precipitation technique. Clinical Chemistry, 30, 1216-1218.

Pleban P.A. and McHugh T. (1983) Copper, magnesium, selenium and zinc in human mononuclear cells and neutrophils. Clinical Chemistry, 29, 1286.

Sunderman F.W., Jr. Personal communication.

Sunderman F.W., Crisostomo C., Reid M.C., Hopfer S.M., Nomoto S. (1984) Rapid analysis of nickel in serum and whole blood by electrothermal atomic absorption spectrophotometry. Annals of Clinical and Laboratory Science, 14, 232-241.

Whitehouse R.C., Prasad A.S., Rabbani P.I. and Cossack Z.T. (1982) Zinc in plasma, neutrophils, lymphocytes, and erythrocytes as determined by flameless atomic absorption spectrometry. Clinical Chemistry, 28, 475-480.

Wills M.R. and Savory J. (1983) Aluminium poisoning: Dialysis encephalopathy, osteomalacia and anaemia. Lancet, ii, 29-34.

THE MEASUREMENT OF NICKEL IN CHRONIC RENAL FAILURE

MICHAEL DRAZNIOWSKY, SUSAN M. CHANNON, IAN S. PARKINSON, MICHAEL K. WARD, TIMOTHY F-H POON* and DAVID N.S. KERR.**
Department of Medicine, Medical School,
Framlington Place, Newcastle upon Tyne, NE2 4HH UK
*Renal Unit, Princess Margaret Hospital,
Kowloon, Hong Kong
**Royal Postgraduate Medical School, Hammersmith Hospi-
Hammersmith Hospital, London, W12 0HS UK

INTRODUCTION

A number of abnormalities in trace element metabolism have been identified in patients with chronic renal failure, both before and after the initiation of regular dialysis treatment. Recently there has been an increasing interest in the possible pathological consequences of such abnormalities.

A number of acute intoxications have been described in patients maintained on haemodialysis. These have been due to trace element contamination of the dialysis fluid, originating from the dialysis equipment or improper water treatment (reviewed by Alfrey and Smythe, 1983). Interest in nickel as a cause of symptoms was stimulated by Webster et al. (1980) who reported that symptoms during haemodialysis including nausea, vomiting, headache, weakness and palpitation were associated with acute nickel intoxication due to the elution of nickel from a nickel-plated stainless steel heater into the water supply used to prepare the dialysis fluid.

We wished to ascertain whether chronic accumulation of nickel from the dialysis fluid was of clinical significance to the haemodialysis population and we have therefore developed a method for the determination of nickel in serum using flameless atomic absorption spectrophotometry.

MATERIALS

Glass distilled water (nickel content <0.7 µg/l) was used for washing and standard preparation. Nitric acid (0.8 M for plastics and 8.0 M for glass) was used to acid-wash containers. A dilute solution of Triton X-100 in distilled water (20 ml/l) (mean nickel content 0.6 µg/l) was used to prepare serum-based standards and to pretreat serum samples. A 5000 µg Ni/l working standard was prepared from a 1 g Ni/l stock solution by dilution with distilled water.

Containers and sampling devices were assessed for contamination by exposure to water, serum or blood for a time in excess of that expected in clinical or laboratory practice.

Acid-washed polypropylene tubes, acid-washed polyethylene bottles, acid-washed glass volumetric flasks, syringes, venepuncture needles, autosampler cups and micropipette tips were found to be free of nickel contamination.

METHODS

In normal subjects, blood samples were obtained by venepuncture using stainless steel needles, while in haemodialysed patients the blood was obtained from the arterial dialysis line.

Blood samples (5 ml) were allowed to clot for 24 h at room temperature in polypropylene tubes. The tubes were centrifuged for 15 min at 850g, the serum decanted into polypropylene tubes and stored at -40 °C until analysis. After thorough vortex mixing, the serum was diluted with an equal volume of Triton X-100 solution to prevent accumulation of carbon in the graphite furnace.

Serum nickel concentrations in normal subjects were determined on a Perkin Elmer 603 atomic absorption spectrophotometer fitted with an HGA-76B graphite furnace and AS-1 autosampler. This was replaced by a Perkin Elmer 3030 spectrophotometer, HGA 500 furnace and AS-40 autosampler for the determination of serum nickel concentrations in patients undergoing haemodialysis.

Instrumental parameters on the 603 spectrophotometer were: wavelength, 232.0 nm; slit setting, 0.2 nm; lamp current, 15 mA; mode, peak area; scale expansion, 0.4; purge gas, argon. The conditions for the 3030 spectrophotometer were similar except: lamp current, 25 mA; expansion, 2.5.

Pyrolytically coated graphite tubes were used as they gave a 2-fold increase in sensitivity over standard graphite tubes. Determinations were performed in triplicate on 50 µl aliquots of samples and standards.

The temperature programme for the 603 spectrophotometer (3030 programme in parenthesis) was: drying stage, 120 °C (90 °C) for 60 s with a ramp of 67 s (60 s); charring stage, 700 °C (1300 °C) for 20 s with a ramp of 76 s (60 s); atomisation stage, 2200 °C for 9 s (5 s); cleaning stage, 2500 °C for 5 s. Attempts were made to reduce the volatility of the analyte and/or eliminate broad band interference from serum by additions of chemical modifiers e.g. ammonium hydroxide, selenium dioxide and inorganic acids. However, these all proved unsuccessful. Therefore a charring temperature of 700 °C (1300 °C) was selected and background correction employed to correct for the broad band interference. During the atomisation stage "maximum power" and "gas stop" were employed.

Aqueous calibration curves were compared to curves obtained by standard additions of nickel to normal and uraemic serum.

In all cases the curves were non-parallel indicating that serum nickel concentrations could not be determined from an aqueous standard curve. Attempts were made to matrix match standards with serum samples but they failed to produce parallel calibration plots with serum. More complete matrix matching of samples and standards was obtained by the use of serum-based standards which gave parallel calibration curves to those of normal and uraemic serum samples.

The detection limit was 0.6 µg/l and the inter-batch coefficient of variation was 5.5% (10 analyses of a 14.4 µg Ni/l sample over 7 months). Nickel concentrations were not significantly different in 31 serum samples analysed on both the 603 and 3030 spectrophotometers.

RESULTS

Statistical calculations were performed using the Mann-Whitney U test unless otherwise stated.

Serum nickel concentrations were determined in 71 normal subjects (39 males, 32 females, age range 19 - 64 years). The median concentration of nickel was 1.0 µg/l, lower quartile 0.6 µg/l, upper quartile 1.4 µg/l, range <0.6 - 3.0 µg/l. The concentration of nickel in males (median 1.0 µg/l) did not differ significantly ($p > 0.05$) from that in females (median 1.1 µg/l).

Serum nickel concentrations were also determined in patients on regular haemodialysis from two hospitals. In 25 patients from Newcastle upon Tyne, UK, significantly increased concentrations ($p < 0.001$) of serum nickel compared with normal subjects were found. Post-dialysis levels (median 8.8 µg/l lower quartile 7.3 µg/l, upper quartile 13.8 µg/l, range 3.0 - 21.4 µg/l) were significantly higher (paired t-test, $p < 0.001$) than pre-dialysis levels (median 8.6 µg/l, lower quartile 5.9 µg/l, upper quartile 12.1 µg/l, range 0.6 - 16.6 µg/l).

In 16 patients from Princess Margaret Hospital, Kowloon, Hong Kong, increased concentrations ($p < 0.001$) of serum nickel compared with normals were again found, but these were significantly lower ($p < 0.001$) than concentrations in Newcastle patients. As with the Newcastle patients, post-dialysis levels (median 3.4 µg/l, lower quartile 3.0 µg/l, upper quartile 3.8 µg/l, range 2.2 - 5.4 µg/l) were significantly higher (paired t-test, $p < 0.01$) than pre-dialysis levels (median 2.9 µg/l, lower quartile 2.4 µg/l, upper quartile 3.4 µg/l, range 1.8 - 4.0 µg/l).

The Newcastle patients had been on dialysis for 1 - 132 months and there was a significant correlation (Spearman's rank correlation coefficient $r_s = 0.62$, $p < 0.01$) between the concentration of serum nickel and the time on dialysis. Initial determination of nickel in tap water, water treated by deionisation and reverse osmosis and dialysis fluid from Newcastle indicated that the highest concentrations were present in the dialysis fluid (2 - 3 µg/l).

CONCLUSION

The method described for the determination of nickel in serum offers advantages over those previously reported (Nomoto and Sunderman, 1970; Andersen et al., 1978; Brown et al., 1981) since chelation and extraction procedures are avoided and low sample volumes are required.

In both Newcastle upon Tyne and Kowloon, serum nickel concentrations were found to be elevated in patients on regular haemodialysis when compared to serum nickel concentrations in normal subjects.

The concentrations of nickel we have detected in patients on regular haemodialysis are considerably lower than those detected by Webster et al. (3000 µg/l). Further studies are needed to establish whether these nickel concentrations correlate with a clinical and/or pathological syndrome as yet unattributed to chronic nickel intoxication.

ACKNOWLEDGEMENTS

This work was supported by the Northern Counties Kidney Research Fund and the Wellcome Trust.

REFERENCES

Alfrey A.C. and Smythe W.R. (1983) In Drukker W., Parsons F.M. and Maher J.F. (Eds.), Replacement of Renal Function by Dialysis, 2nd Edition, pp. 804-810. Martinus Nijhoff, Boston.

Andersen I., Torjussen W. and Zachariasen H. (1978) Clinical Chemistry, 24, 1198-1202.

Brown S.S., Nomoto S., Stoeppler M. and Sunderman F.W. Jr. (1981) Clinical Biochemistry, 14, 295-299.

Nomoto S. and Sunderman F.W. Jr. (1970) Clinical Chemistry, 16, 477-485.

Webster J.D., Parker T.F., Alfrey A.C., Smythe W.R., Kubo H., Neal G. and Hull A.R. (1980) Annals of Internal Medicine, 92, 631-633.

461637

SERUM NICKEL CONCENTRATIONS IN PATIENTS WITH UNSTABLE ANGINA PECTORIS AND ACUTE MYOCARDIAL INFARCTION

CHARLES N. LEACH Jr.,* JEANNE V. LINDEN,** SIDNEY M. HOPFER,** CHRISTINA CRISTOSTOMO,** and F. WILLIAM SUNDERMAN Jr.**
*Department of Medicine, New Britain General Hospital, New Britain, CT 06050, and **Department of Laboratory Medicine, University of Connecticut Medical School, Farmington, CT 06032, USA

INTRODUCTION

Using emission spectrographic analysis, D'Alonzo and Pell (1963) observed elevated Ni concentrations in 19 of 20 serums obtained from patients with acute myocardial infarction within 24 hours after admission to the hospital. Using flame atomic absorption spectrometry, Sunderman et al. (1970) and McNeely et al. (1971) observed increased Ni concentrations in 21 of 29 serums obtained from patients with acute myocardial infarction in the interval from 12 to 36 hours after onset of symptoms and in 6 of 22 serums from patients with acute myocardial ischemia without infarction. Emission spectrographic analyses performed by Tarala (1970), Nozdryukhina (1978), and Howard (1980) confirmed that hypernickelemia occurs in most patients with acute myocardial infarction. Sunderman et al. (1984) developed a sensitive and precise method for analysis of serum nickel by direct electrothermal atomic absorption spectrometry with Zeeman background correction. The present paper describes application of this new technique for measurement of Ni concentrations in serums from healthy subjects, from patients with acute myocardial infarction, and from patients with unstable angina pectoris. 21

MATERIALS AND METHODS

Blood specimens were obtained by use of polyethylene iv cannulae and acid-washed syringes from (A) 30 healthy subjects, age 23 to 84 years, (B) 46 patients with acute myocardial infarction, and (C) 29 patients admitted to the cardiac intensive care unit with unstable angina pectoris. In groups B and C, blood samples were collected on admission, at 8 hour intervals on day 1, and daily thereafter up to 3 days. Ni concentrations in serum and whole blood were measured by electrothermal atomic absorption spectrometry (Sunderman et al. 1984).

RESULTS

Ni concentrations in serum and whole blood from 30 healthy subjects ranged from <0.1 to 1.1 µg/L (Table 1). Hypernickelemia (serum Ni >1.1 µg/L) was observed in 36 of 46 patients with acute myocardial infarction during 72 hours after admission to the cardiac intensive care unit (Table 2). The peak Ni concentrations averaged 4.5 ± 10.2 µg/L (range = 0.7 - 21.2 µg/L). For comparison, hypernickelemia was observed in 16 of 29 patients with unstable angina pectoris during the same interval (Table 3). The peak Ni concentrations averaged 1.5 ± 1.4 µg/L (range = 0.4 - 3.1 µg/L). Ni concentrations in 50 specimens of whole blood from patients with myocardial infarction or unstable angina pectoris were closely correlated with corresponding Ni concentrations in serum (correlation coefficient = 0.82, P <0.001); the following least-squares regression equation was obtained: $Y = 0.53X + 0.21$, where Y denotes the whole blood Ni concentration and X denotes the serum Ni concentration. The peak concentrations of serum Ni in patients with acute myocardial infarction were not correlated in magnitude or in timing with the peak activities of serum creatine kinase, including isoenzyme fractions. Hypernickelemia was observed in three patients with myocardial infarction who were treated with streptokinase. No other associations between hypernickelemia and medications were found. Serum Ni concentrations were unrelated to age, sex, time of day, cigarette smoking, or clinical outcome. When seven patients with hypernickelemia following myocardial infarction were retested 1 to 6 months later, the serum Ni concentrations were all < 1.1 µg/L.

TABLE 1. Nickel concentrations in serum and whole blood of healthy adult residents of Connecticut

Healthy subjects	Age range (years)	No.	Ni concentration (µg/L) (mean ± SD) serum	whole blood
men:	25-72	15	0.23 ± 0.20	0.25 ± 0.20
women:	23-84	15	0.33 ± 0.27	0.38 ± 0.34
all subjects:	23-84	30	0.28 ± 0.24	0.34 ± 0.28
range:			(<0.1 - 1.1)	(<0.1 - 1.1)

TABLE 2. Serum Ni concentrations in 46 patients with acute myocardial infarction

Hours after admission	No.	Serum nickel concentrations (µg/L) mean ± SD	range	Serum Ni conc. >1.1 µg/L[a]
0-8	19	0.5 ± 0.7	<0.1 - 2.8	16%
9-16	32	1.0 ± 1.0[b]	<0.1 - 5.5	22%
17-24	27	2.9 ± 4.5[b]	0.2 - 21.2	59%
25-48	36	1.8 ± 1.9[b]	0.1 - 8.3	55%
49-72	35	1.6 ± 1.4[b]	0.1 - 7.2	49%

[a] Peak serum Ni concentrations were >1.1 µg/L in 36 of 46 patients (78%) during 72 h after admission.
[b] $P < 0.001$ versus healthy controls.

TABLE 3. Serum Ni concentrations in 29 patients admitted with unstable angina pectoris

Hours after admission	No. of subjects tested	Serum Ni concentration (µg/L) mean ± SD	range	Serum Ni conc. >1.1 µg/L[a]
0 - 8	19	0.9 ± 1.6	0.1 - 6.7	11%
9 - 16	18	0.8 ± 0.6[b]	0.1 - 1.9	39%
17 - 24	16	0.9 ± 0.7[b]	0.1 - 3.0	19%
25 - 48	16	1.1 ± 0.9[b]	0.1 - 3.0	31%
49 - 72	11	1.0 ± 0.6[b]	0.1 - 1.7	45%

[a] Peak serum Ni concentrations were > 1.1 µg/L in 16 of 29 patients (55%) during 72 h after admission
[b] $P < 0.01$ versus healthy controls.

DISCUSSION

This study shows that hypernickelemia develops in most patients with acute myocardial infarction and in approximately one-half of patients with unstable angina pectoris, based upon analyses by a sensitive, precise, and specific technique. These determinations confirm the qualitative findings of prior investigators, even though the new analytical technique yields quantitative measurements of serum Ni concentrations that are substantially lower than previously reported. Coronary spasm is believed to play an important etiologic role in unstable angina and acute myocardial infarction. Although Ni has not been documented as a factor in human coronary spasm, Rubanyi and Kovach

(1980) and Rubanyi et al. (1984) have summarized evidence that low concentrations of exogenous $NiCl_2$ induce coronary vasoconstriction in isolated perfused dog heart, dog heart *in situ*, and isolated canine coronary arterial strips. Experiments on myocardium of anesthetized dogs showed that low doses of exogenous $NiCl_2$ inhibit post-occlusion reactive hyperemia and arterial hypoxemia. $NiCl_2$ induces coronary vasoconstriction in the presence of the coronary vasodilatory mechanisms that are triggered by hypoperfusion, hypoxia, and adenosine infusion. Based upon these findings, Rubanyi et al. (1984) speculated that endogenous Ni release may play a role in ischemic coronary vasoconstriction. Our data support the hypothesis that hypernickelemia contributes to the pathogenesis of myocardial infarction for the following reasons: (A) Hypernickelemia frequently accompanies unstable angina pectoris and acute myocardial infarction, but is more pronounced and sustained in patients with myocardial infarction. (B) Two patients who suffered extensions of their infarcts had persistently elevated concentrations of serum Ni in 5 of 5 and 8 of 8 samples, respectively. These elevations persisted after the initial episode and antedated the second, suggesting a causal relationship. (C) Peak serum Ni concentrations did not correlate with peak creatine kinase activities, and occurred in varying temporal relationships to serum creatine kinase elevations in patients with myocardial infarction. The mechanism and source of nickel release in patients with myocardial infarction and unstable angina pectoris remains to be elucidated.

REFERENCES

D'Alonzo C.A. and Pell S. (1963) Arch. Environ. Health, 6, 381-385.

Howard J.J.H. (1980) Clin. Chem. 26, 1515.

McNeely M.D., Sunderman F.W. Jr., Nechay M.W. and Levine H. (1971) Clin. Chem. 17, 1123-1128.

Nozdryukhina L.R. (1978) in Kirchgessner M. (Ed.) Trace Element Metabolism in Man and Animals, pp. 336-339. Technische Universitat Munchen, Freising-Weihenstephan.

Rubanyi G. and Kovach A.G.B. (1980) Acta Physiol. Acad. Sci. Hung., 55, 345-353.

Rubanyi G., Ligeti L., Koller A. and Kovach A.G.B. (1984) Mol. Cell Cardiol., 16, 533-546.

Sunderman F.W. Jr., Nomoto S., Pradhan A.M., Levine H., Bernstein S.H. and Hirsch R. (1970) New Engl. J. Med., 283, 896-899.

Sunderman F.W. Jr., Crisostomo C., Reid M.C., Hopfer S.M. and Nomoto S. (1984) Ann. Clin. Lab. Sci., 14, 232-241.

Tarala G.T. (1970) Kardiologia, 10, 146-147.

461638

URINARY EXCRETION OF NICKEL AND CHROMIUM IN WORKERS OF A STEEL FOUNDRY

Antero Aitio*, Antti Tossavainen*, Tom Gustafsson*, Mirja Kiilunen*, Kauko Haapa**, and Jorma Järvisalo*
* Institute of Occupational Health
SF-00290 Helsinki, Finland
**A. Ahlström Oy, SF-48600 Karhula, Finland

INTRODUCTION

The relationship between concentrations of nickel in the air, and in biological specimens is closest in exposure to water soluble nickel salts ~~(Aitio 1984)~~. The purpose of this study was to extend the knowledge on the biological monitoring of nickel and chromium in cast cleaning in a steel foundry. The characteristics of the airborne dust were also investigated.

SUBJECTS, MATERIALS AND METHODS

Four flame cutters and 2 arc welders, who had been working in the same plant for several years, were studied. On average, the alloys worked on contained 4 to 11 % nickel and 13 to 23 % chromium. On Wednesday, Thursday and Friday, dust specimens and all the urine excreted (in 5 portions/day) were collected. In addition, two daily spot urine specimens were collected on four days before, and three days after the study period.

Air sampling and analysis. Fume samples were collected on cellulose ester membrane filters (Millipore AAWP037, 0.8 μm pore size) placed inside the welding mask. The flow rate averaged 2 L/min and each day two 4-h samples were taken. At the same time, several short term samples for microscopic analysis were collected on polycarbonate filters (Nuclepore, 0.2 μm pore size) during different work phases.

The amount of total dust was measured gravimetrically. For metal analysis, the fume samples were treated successively with 5 ml of concentrated nitric and hydrochloric acid, evaporated to dryness, dissolved in 5 per cent HNO_3 and filtered through polycarbonate filters (Nuclepore). Metal concentrations of the solutions were measured with flame AAS.

The samples collected directly on Nuclepore filters as well as the insoluble residues were studied with scanning electron microscopy (JEOL-100CX, AS1D4) and X-ray microanalysis(PGT-III-system) after covering with a gold layer (about 20 nm in thickness) in an ion-sputtering device(JEOL Fine Coating, JFC 1100). The X-ray analyser was linked to a microcomputer (Luxor ABC-80L) for data storage and calculation. The detection limit for chromium and nickel was estimated to be 3 and 10 %, respectively.

Biological specimens Urine specimens were collected in prewashed polyethylene bottles. The workers were instructed on how to avoid contamination. The analyses of urinary and plasma nickel and chromium were performed using a direct dilution method of acidified urine/plasma with a Perkin Elmer 4000 AA spectrometer (HGA 400 graphite furnace) (Kiilunen et al., submitted). The day-to-day variation of the analyses was 5.6 and 7.0 % for nickel and chromium, at levels of 0.55 and 0.46 μmol/L levels, respectively.

RESULTS

The concentration of nickel in the air varied from 50 to 1100 μg/m^3, that of acid soluble chromium from 56 to 790 μg/m^3 (Table 1).

Electron microscopic studies showed submicron fume chain agglomerates together with a large amount of spherical particles 1 to 7 μm in diameter (Fig 1). Four percent of these particles were highly enriched with chromium, while some particles contained more than 50% nickel, but no chromium. After acid dissolution of the fumes, nickel rich particles could be detected no more, whereas the chromium rich particles had an etched appearance .

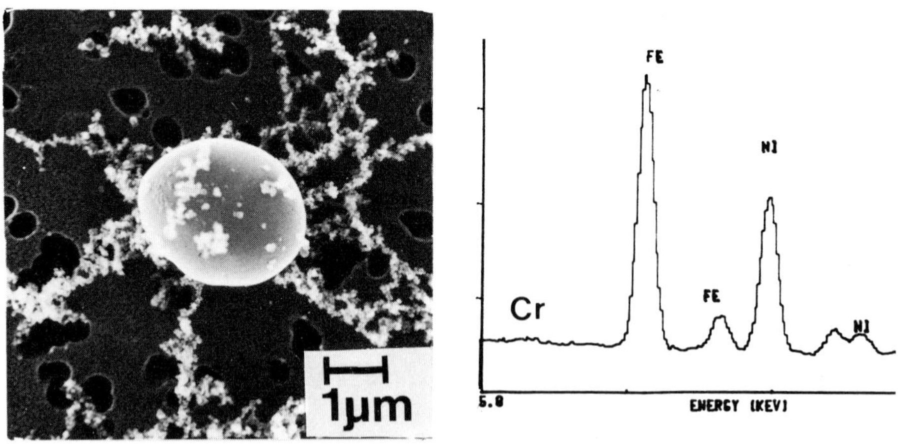

Fig 1. Secondary electron image and X-ray spectrum of a nickel-rich particle in flame cutting fume.

The concentrations of nickel and chromium were high in the urine: the averages for the six workers were between 0.45 and 1.76 µmol/L for chromium, and between 0.30 and 1.30 µmol/L for nickel (Table 1). Exposure-related variation of nickel or chromium concentrations was either minor or could not be detected at all (Fig. 2). No relationship could be established between the concentrations of the metals in the air and urine. Two of the workers showed a tendency for increasing urinary nickel and chromium concentrations during the working day. Five workers showed an increasing concentration of both chromium and nickel in the afternoon urine specimens during the working week. The decrease during the week-end did not reach statistical significance; during a 4-week vacation the urinary nickel and chromium concentrations showed a 33% and 29% decrease on average. The pattern of excretion of the two metals was similar in the welders and in the flame cutters.

The variations in the metal concentrations in urine within an individual were small, the relative standard deviation being 22-29 % for chromium, and 16-25 % for nickel in the six workers. The molar Cr/Ni ratio was 0.70 to 1.55 in the air, 2.8 to 10.7 in the serum, and 1.2 to 2.2 in the urine.

Table 1. The three-day mean concentrations of nickel and chromium in the breathing zone and urine of four flame cutters and two welders from a steel foundry.

	Airborne concentration			Concentration in urine	
	Nickel	Chromium	Total dust	Ni	Cr
	µg/m^3	µg/m^3	mg/m^3	µmol/L	µmol/L
Flame cutter I	1100	730	37.0	1.30	1.76
Flame cutter II	248	153	12.2	0.60	1.45
Flame cutter III	580	790	17.3	0.30	0.52
Flame cutter IV	145	103	5.4	0.45	0.45
Welder I	50	56	3.4	0.39	0.90
Welder II	152	210	8.3	0.57	1.15

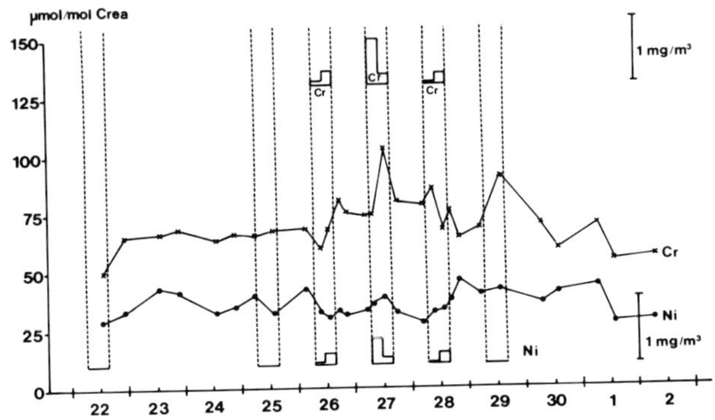

Fig 2. Urinary excretion of nickel and chromium in a flame cutter.

DISCUSSION AND CONCLUSIONS

In the cleaning of acid resistant steel castings, there was practically no relationship between actual exposure and urinary concentrations of nickel or chromium. However, remarkably high concentrations of both metals were excreted in the urine, and these concentrations were rather stable within a single worker. They showed a slow increase during the working week and their disappearance was slow. The urinary excretion of nickel and chromium therefore probably reflects cumulative rather than short-term exposure. In contrast to previous work (Tola et al. 1977; Rahkonen et al. 1983) on urine nickel and chromium no daily exposure-related fluctuations were seen in this study; it is evident that in these circumstances, even the welders are mostly exposed to rather insoluble compounds. Nickel and chromium seem to segregate in the flamecutting process; this results in two different types of particles; those enriched with chromium, and those with nickel.

REFERENCES

Aitio A. (1984) In Nickel in the Human Environment, IARC Sci. Publ. Ser, in press.

Rahkonen E., Junttila M.-L., Kalliomäki P.-L., Olkinuora M., Koponen M. and Kalliomäki K. (1983) Int. Arch. Occup. Environ. Health, 52, 243-255.

Tola S., Kilpiö J., Virtamo M., Haapa K. (1977) Scand. J. Work Environ. Health, 3, 192-202.

461639

EXTERNAL AND INTERNAL NICKEL EXPOSURE OF ARC WELDERS DURING ONE WORKING WEEK

KARL-HEINZ SCHALLER, ANDREAS ZOBER and HELMUT VALENTIN

Institute of Occupational and Social Medicine,
University Erlangen-Nürnberg,
Schillerstr. 25, D-8520,
Erlangen, FRG

INTRODUCTION

Among the various groups who are occupationally exposed to nickel, welders who work with chromium-nickel containing filler metals are numerically important. In the FRG the medical surveillance of such welders is subject to the guidelines governing medical surveillance at work G 39 (welding fumes), resp. G 15 (Cr-VI compounds) and G 38 (Ni or its compounds) including a biological monitoring programme. Already in 1982 we have extensively reported the results of biological monitoring with respect to arc welders (Zober 1982). The type of investigation selected (cross sectional study) however made it impracticable to provide reliable data for the relationships between external and internal exposure. It was equally difficult to say anything about the intraindividual variability of internal exposure. The results of a new study along these lines are reported now.

GROUPS AND METHODS

We examined twenty arc welders from two factories with a mean age of 39.5 (19 to 57) years, who have been working as welders for an average of 19 (3 to 42) years. 70 % were smokers, 20 % ex-smokers and only 10 % non-smokers. All these welders had been working with Cr-Ni containing filler metals for 50 % (20 to 100 %) of their working hours for an average of 13 (1 to 27) years. One shielded arc welder had worked exclusively on structural steel and therefore had suffered only subsidiary exposure. The Ni-content of the electrodes was 8-10 %, in special cases 65 %. In order to assess the external and internal exposure level on an objective and quantitative basis, measurements were taken of the welding-fume concentration in the breathing zone behind the shield, in addition to which urine (Ni-U) and plasma (Ni-P) analyses were carried out. The air samples were taken on a daily basis over a period of roughly two hours per welder. The urine samples were taken before, during and after each shift in the course of a working week and on the following Monday; the blood samples were taken before and after the shift only. The air samples were analysed by gravimetry and ET-AAS after wet digestion with perchloric, sulfuric and nitric acids (Moreton et al., 1983). The Ni concentrations in the biological samples were determined by elecrothermal AAS.

RESULTS AND DISCUSSION

The values for the external exposure levels as measured for the total group of subjects during one week's work using Cr-Ni containing filler metals with a 20 to 40 % effective arc-burning time - this work essentially being done in workshops with controlled ventilation - are set out in Table 1.

TABLE 1. External welding fume exposure (medians and 90 % ranges)

	Total Group	Group 1	Group 2	Group 3	Exp. Limits
Resp.dust (mg/m^3)	5.3 (0.4/53)	2.1 (0.3/14)	5.6 (0.3/22)	16.7 (0.3/101)	6
Nickel ($\mu g/m^3$)	5.8 (0.6/78)	1.2 (1.2/14)	4.6 (1.2/21)	13.9 (1.2/656)	500

Depending on the extent to which coated filler metals were used, the rates of exposure fall into three different groups:

a) group 1 - predominantly tungsten inert-gas and metal inert-gas welding, with up to 20 % manual arc welding using coated stick electrodes,

b) group 2 - occasional tungsten inert-gas and metal inert-gas welding, with over 20 to 60 % manual arc welding using coated stick electrodes,

c) group 3 - predominantly (over 60 to 100 %) manual arc welding using coated stick electrodes.

The results obtained on the basis of this classification are summarized in Table 1. None of the measurements taken during the metal inert-gas and tungsten inert-gas welding operations was in excess of the general dust limit of 6 mg/m^3.

The Ni readings only reached the recommended TRK-value of 500 $\mu g/m^3$ in two cases when Ni-based alloys were being welded, and on one occasion this value was exceeded by a factor of 2 when using a pure Ni-electrode.

Internal exposure levels with respect to the total group of subjects over a period under analyses are presented in Table 2. The substantial variations in the external exposure level - increasing from exposure groups 1 to 3 along with the extent to which coated stick electrodes were used - are reflected in the results for the corresponding biological parameters.

TABLE 2. Internal Ni exposure (medians, 90 % ranges)

	Total Group	Group 1	Group 2	Group 3	Exp. Limits
Ni-U (µg/g cr.)	2.5 (0.4/6.7)	1.7 (0.8/7)	2.4 (0.2/8)	3.0 (1.2/10)	25
Ni-P (µg/l)	0.4 (0.4/5)	0.4 (0.4/2)	0.4 (0.4/5)	1.3 (0.4/5)	10

The renal excretion of 25 µg/g of creatinine currently under discussion as a tolerable action level in the FRG was not attained even once. The median Ni-plasma levels are not different from those of non-affected persons. In overall terms, it is evident that welding operations using filler metals with a Ni content of 8 to 10 % in workshops with controlled ventilation results in low external and low internal Ni exposure.

The kinetics of Ni for the entire period of examinations may be assessed from the biological concentrations in the course of the day and the week. None of the exposure groups's median values showed any significant increase in renal elimination over the day. Fig. 1 demonstrates the behaviour of the Ni excretion in urine during the day and the week for the high-exposure group 3.

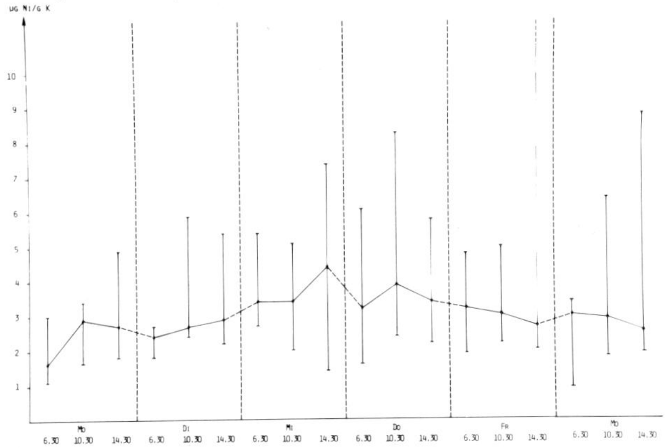

Fig.1. Kinetics of Ni in urine (medians, 66 % ranges) (Group 3)

Neither the data from the urine nor the plasma analysis recorded any significant increase in basal excretion by way of an accumulation as the week progressed, and there was no substantial rise in these values in the course of a day either. Only for the exposure-free interval "weekend" could a decline in the renal Ni elimination be observed for four subjects.

In view of the absolute internal exposure level, the half-line of renal excretion could only be calculated with two of the test subjects. The resulting half-line value was around 53 hours. The relationship between the Ni levels in urine and plasma could be described by a linear regression. The coefficient of correlation is 0.44 for 237 pairs of values. It was not possible to characterise any relationships between external and internal exposure.

CONCLUSIONS

Within the scope of the present study, not only Ni analyses were carried out, but also the external and internal Cr exposures were measured. Detailed results will be presented elsewhere, ~~(Zober et al., 1984)~~. Therefore the conclusion for Ni can only be extrapolated from the results for chromium:

1. In order to assess the external exposure to welding fumes, measurements should be taken in the breathing zone behind the shield over a sufficiently long period (more than 2 h). In the case of metal inert-gas and tungsten inert-gas welding, the external exposure level is insignificant from the point of view of industrial health. Manual arc welding- and, within this context, surface welding in particular - produces readings significantly in excess of the limit values. When filler metals with 18 to 20 % Cr and 8 to 10 % Ni are used, however, these limits are only exceeded by Cr and not by Ni.
2. Urine analyses are adequate as a basis for quantifying the internal exposure level in practice. There is no need to conduct plasma analyses, except in conditions of high exposure such as those encountered in welding operations using pure Ni electrodes.

REFERENCES

Moreton J., Bettelley J., Mathers H., Nicholls A., Perry R. W., Ratcliffe D.B. and Svensson L. (1983) Ann.Occup. Hyg. 27, 2, 137-156.

Zober A. (1982) Forschungsbericht Nr. 317, Hrsg. Bundesanstalt f. Arbeitsschutz, Dortmund, Verlag f. neue Wissenschaft GmbH, Bremerhaven.

Zober A., Weltle D. and Schaller K.H. (1984) Schweißen und Schneiden (in press).

EFFECTS OF $NiCl_2$ AND DIETHYLDITHIOCARBAMATE ON HEME OXYGENASE, METALLOTHIONEIN, GLUTATHIONE, AND TRACE METALS IN RAT LIVER AND KIDNEY

F. WILLIAM SUNDERMAN Jr.
Departments of Laboratory Medicine and Pharmacology, University of Connecticut School of Medicine, Farmington Connecticut 06032, USA

INTRODUCTION

Maines and Kappas (1977) reported that sc administration of $NiCl_2$ to rats causes marked induction of renal heme oxygenase activity. Sunderman et al. (1983b) confirmed this finding and delineated the time-course, dose-effect, organ-selectivity, and species-specificity relationships. The present paper summarizes the synergistic effect of diethyldithiocarbamate and $NiCl_2$ on microsomal heme oxygenase activity, and the influence of these compounds, singly or in combination, on metallothionein, glutathione, and trace metal concentrations in rat liver and/or kidney. The goal of this paper is to consolidate and interpret data that have been reported in several recent papers from our laboratory (Sunderman et al., 1982, 1983a,b, 1984).

MATERIALS AND METHODS

The test substances were sodium diethyldithiocarbamate (DDC) and ultrapure $NiCl_2$. The experimental animals were male rats of the Fischer-344 strain (170-300 g, body wt.), maintained in plastic cages with a diet of Purina rat chow and tap water, and fasted for 17 h prior to sacrifice. The analytical methods were (a) heme oxygenase assay in washed microsomes, based upon enzymic release of CO from the α-methene bridge of heme and quantitation of CO by gas chromatography (Sunderman et al., 1982); (b) metallothionein determination in cytosol by the Cd-saturation assay of Onosaka and Cherian (1982), with Cd analysis by atomic absorption spectrometry; (c) assay of reduced glutathione (GSH) and oxidized glutathione (GSSG) in tissue homogenates by the kinetic method of Tietze (1969), based upon complexation of GSH with N-ethylmaleimide and chromogenic reaction with Ellman's reagent (dithiobis(nitrobenzoic acid)); and (d) quantitation of Ni, Cu, and Zn in tissue homogenates by acid digestion and plasma emission spectrometry (Sunderman et al., 1984).

RESULTS

The synergistic action of DDC and $NiCl_2$ on microsomal heme oxygenase activity is illustrated in Table 1. In the rats that received DDC and $NiCl_2$ at the highest dosages, heme oxygenase activity was increased 11-fold in liver and 16-fold in kidney. As summarized in Table 2, $NiCl_2$ treatment increased Ni and Zn concentrations in liver and kidney; DDC treatment increased Zn concentration in liver and decreased Cu concentrations in kidney. Combined administration of DDC

Table 1. Effects of DDC and $NiCl_2$ on heme oxygenase activity.

DDC dose (mmol/kg, im, 17.2 h)	$NiCl_2$ dose (mmol/kg, sc, 17 h)	No. of rats	Heme oxygenase activity (nmol/h/mg protein, mn±SD) liver	kidney
0	0	5	21 ± 4	10 ± 2
0	0.25	5	26 ± 5	58 ± 21[a]
67	0	5	39 ± 5[a]	12 ± 3
0.67	0.25	5	193 ± 23[b]	117 ± 5[b]
1.33	0	5	65 ± 7[a]	21 ± 5[a]
1.33	0.25	5	221 ± 37[b]	163 ± 36[b]

[a] $p < 0.05$ vs controls. [b] $p < 0.01$ vs DDC or $NiCl_2$ alone.

Table 2. Effects of DDC and $NiCl_2$ on metal concentrations.

DDC dose (mmol/kg, im, 17.2 h)	$NiCl_2$ dose (mmol/kg, sc, 17 h)	No. of rats	Metal concentrations (nmol/g wet wt., mn ± SD) Ni	Cu	Zn
liver					
0	0	8	< 20	64 ± 8	442 ± 32
0	0.25	4	53 ± 26[a]	69 ± 9	590 ± 62[a]
1.33	0	8	< 20	60 ± 8	622 ± 63[a]
1.33	0.25	4	308 ± 63[b]	53 ± 4[a]	613 ± 54[a]
kidney					
0	0	8	< 20	93 ± 12	302 ± 25
0	0.25	4	656 ± 94[a]	93 ± 13	459 ± 39[a]
1.33	0	8	< 20	68 ± 6[a]	288 ± 30
1.33	0.25	4	689 ± 98[a]	59 ± 3[a]	294 ± 6[b]

[a] $p < .05$ vs controls. [b] $p < 0.05$ vs $NiCl_2$.

Table 3. Effects of DDC and $NiCl_2$ on metallothionein conc.

DDC dose (mmol/kg, im, 17.2 h)	$NiCl_2$ dose (mmol/kg, sc, 17 h)	No. of rats	Metallothionein (μg/g wet wt., mn ± SD) liver	kidney
0	0	6	39 ± 10	129 ± 6
0	0.25	6	156 ± 38[a]	239 ± 49[a]
1.33	0	5	296 ± 55[a]	113 ± 28
1.33	0.25	7	356 ± 37[b]	208 ± 19[a]

[a] $p < 0.001$ vs controls. [b] $p < 0.05$ vs DDC or $NiCl_2$ alone.

Table 4. Effects of DDC and $NiCl_2$ on hepatic glutathione.

DDC dose (mmol/kg, im, 17.2 h)	$NiCl_2$ dose (mmol/kg, sc, 17 h)	No. of rats	Glutathione conc. (mmol/q, wet wt., mn±SD) GSH	GSSG
0	0	15	4.5 ± 0.5	0.06 ± 0.02
0	0.25	8	4.5 ± 0.5	0.06 ± 0.02
0.67	0	5	5.0 ± 1.2	0.05 ± 0.01
0.67	0.25	5	6.5 ± 1.2[a]	0.05 ± 0.01
1.33	0	7	5.0 ± 0.6	0.05 ± 0.01
1.33	0.25	5	6.7 ± 1.3[a]	0.06 ± 0.01

[a] $p < 0.05$ vs DDC or $NiCl_2$ alone.

and $NiCl_2$ markedly enhanced Ni concentrations in liver, but not in kidney, and caused slightly diminished Cu concentrations in both organs. Effects of DDC and $NiCl_2$ on metallothionein concentrations are given in Table 3. $NiCl_2$ caused 4-fold increase of metallothionein concentration in liver and 2-fold in kidney; DDC caused 7.5-fold induction of metallothionein concentration in liver, but did not affect the concentration in kidney. Combined administration of DDC and $NiCl_2$ caused further increase of metallothionein concentration in liver, but not in kidney. Table 4 indicates that neither DDC nor $NiCl_2$, given alone to rats at the specified dosages, significantly affected hepatic GSH or GSSG concentrations. Combined treatment with DDC and $NiCl_2$ caused synergistic increase of hepatic GSH concentration without affecting hepatic GSSG concentration.

DISCUSSION

Kikuchi and Yoshida (1983) showed that two distinct metabolic pathways are involved in the induction of heme oxygenase activity - one, heme-mediated, and the other, metal mediated; both pathways involve increased functioning of levels of mRNA for heme oxygenase. The present finding suggests that the synergistic effects of DDC and $NiCl_2$ on renal heme oxygenase activity are mediated primarily by an increased intracellular heme pool, as a consequence of enhanced renal degradation of cytochrome P-450 (Masuda and Nakayama 1983). This inference is supported by the lack of synergistic actions of DDC and $NiCl_2$ on the concentrations of Ni, Zn, or metallothionein in rat kidney. On the other hand, the synergistic effects of DDC and $NiCl_2$ on hepatic heme oxygenase activity are apparently mediated by enhanced hepatic uptake of Ni as a Ni-DDC complex. This inference is supported by the actions of DDC and $NiCl_2$ on the concentrations of Ni and metallothionein in rat liver. The synergistic effect of DDC and $NiCl_2$ on hepatic GSH concentration may reflect enhanced hepatic uptake of DDC as a Ni-DDC complex. The increased intracellular concentration of DDC probably blocks glutathione metabolism by inhibiting γ-glutamyl transferase and glutathione S-transferase activities (Dierickx 1984; Maggiore et al. 1982).

REFERENCES

Dierickx P.J. (1984) Pharmacol. Res. Commun., 16, 135-143.

Kikuchi G. and Yoshida T. (1983) Mol. Cell. Biochem., 53, 163-183.

Maggiore A., Ficarra M.G., Savarese N. and Orecchio F. (1982) Quad. Sclavo. Diagn., 18, 293-301.

Maines M.D. and Kappas, A. (1977) in Brown S.S. (Ed.), Clinical Chemistry and Chemical Toxicology of Metals, pp. 75-81, Elsevier, Amsterdam.

Masuda Y. and Nakayama N. (1983) Biochem. Pharmacol., 32, 3127-3135.

Onosaka S. and Cherian M.G. (1982) Toxicol. Appl. Pharmacol. 63, 270-274.

Sunderman F.W. Jr., Bibeau L.M. and Reid M.C. (1983), Toxicol. Appl. Pharmacol., 71, 436-444.

Sunderman F.W. Jr., Downs J.R., Reid M.C. and Bibeau L.M. (1982) Clin. Chem. 28, 2026-2032.

Sunderman F.W. Jr. and Fraser C.B. (1983) Ann. Clin. Lab. Sci., 13, 489-495.

Sunderman F.W. Jr., Reid M.C., Bibeau L.M. and Linden J.V. (1983) Toxicol. Appl. Pharmacol., 68, 87-95.

Sunderman F.W. Jr., Zaharia O., Reid M.C., Belliveau J.F., O'Leary G.P. and Griffin H. (1984) Toxicology, 32, 11-21.

Tietze F. (1969) Anal. Biochem., 27, 502-522.

THE EFFECT OF NICKEL AND MANGANESE ON NATURAL KILLER CELL ACTIVITY

RALPH J. SMIALOWICZ,
Developmental Biology Division, Health Effects Research Laboratory, Research Triangle Park,
North Carolina 27711, USA

INTRODUCTION

Nickel (Ni) is an example of a toxic heavy metal of environmental concern which has been found to impair various components of the immune system (Smialowicz et al., 1984a) and to suppress host resistance to infection (Adkins et al., 1979). Certain Ni compounds have also been shown to be carcinogenic in man and animals (Sunderman, 1981).

During the course of our research on the effects of Ni on the immune system of mice, we discovered that natural killer (NK) cells were particularly sensitive to suppression by $NiCl_2$ (Smialowicz et al., 1984b). NK cells have been receiving increased attention in recent years because of their potential role as primary cytotoxic effector cells against tumors. These lymphocytes have been shown to be *in vivo* effectors against transplanted tumors and have been suggested to play a role against newly arising tumors (Herberman and Ortaldo, 1981).

While Ni suppresses murine NK activity, manganese (Mn) in the form of $MnCl_2$, enhances NK activity (Rogers et al., 1983 and Smialowicz et al., 1984c). The observation that Ni suppresses NK activity while Mn enhances it takes on added significance because of the results of Sunderman and McCully (1983) who showed that Mn dust inhibits the carcinogenicity of nickel subsulfide.

This paper summarizes the results of studies on the NK activity of mice injected with either $NiCl_2$ or $MnCl_2$.

METHODS AND RESULTS

Mice of several different strains were given a single intramuscular injection in the thigh on a body weight basis 24 hr prior to assay. Mice received a dose of 18.3 mg/kg

Ni as $NiCl_2$ or 40 or 80 mg/kg Mn as $MnCl_2$ in saline. Saline injected mice served as controls. The Ni and Mn concentrations were determined by neutron activation analysis.

NK cell activity was determined using three different assay systems. These included the following: 1) an in vitro 4 hr ^{51}Cr-release assay in which spleen cells were cocultured with ^{51}Cr-labeled YAC-1 tumor target cells; 2) an in vivo clearance assay in which the clearance of tumor cells from the lungs of mice was determined 4 hr after an intravenous injection of ^{125}I-labeled YAC-1 cells; and 3) an in vivo tumor assay in which the number of tumor foci in the lungs was determined 2-3 weeks after an intravenous injection of B16-F10 melanoma cells.

TABLE 1. Comparison of the effects of Ni and Mn on NK activity.

Treatment[a]	$\bar{x}$% ^{51}Cr Release±SE[b]	$\bar{x}$% ^{125}I Recovered[c]	$\bar{x}$ Tumors/ Lung±SE[d]
Saline	14.9 ± 1.6	0.44	6.7 ± 1.3
$NiCl_2$	7.1 ± 1.6[e]	1.15[e]	63.8 ± 7.6[e]
Saline	12.5 ± 1.3	0.71	47.5 ± 10.1
$MnCl_2$	30.5 ± 2.0[e]	0.45[e]	9.0 ± 1.5[e]
Saline	12.3 ± 0.8	-	-
$NiCl_2$	4.8 ± 1.1[e]	-	-
$MnCl_2$	21.9 ± 1.7[e]	-	-
$MnCl_2$+$NiCl_2$	23.4 ± 0.8[e]	-	-

[a] Groups of 8-10 mice were given a single intramuscular injection of 18.3 mg/kg $NiCl_2$ or 40 or 80 mg/kg $MnCl_2$.

[b] Four hour in vitro NK activity assay against YAC-1 cells.

[c] Four hour in vivo tumor cell clearance assay expressed as the geometric mean.

[d] Mice were injected intramuscularly with saline, $NiCl_2$ or $MnCl_2$ 24 hr before an intravenous injection of 10^5 B16-F10 melanoma cells.

[e] P<0.01, metal vs saline.

Table 1 summarizes representative results from several experiments in which the effects of Ni and Mn on NK activity were determined. A single intramuscular injection of $NiCl_2$ (18.3 mg/kg) caused a significant suppression of NK cell activity as determined by the in vitro ^{51}Cr-release assay. The clearance of radiolabeled tumor cells from the lungs of Ni-treated mice was also significantly reduced compared with saline controls. A single injection of 18.3 mg/kg $NiCl_2$ 24 hr before injection of B16-F10 melanoma cells significantly increased the number of resulting lung tumors compared with saline controls.

In contrast, mice injected with $MnCl_2$ displayed enhanced NK activity as measured by the in vitro ^{51}Cr release assay. The enhancement of NK activity by Mn was found to be associated with increased serum levels of interferon. Furthermore, treatment of Mn-injected mice with rabbit anti-interferon α, β but not γ completely abrogated NK augmentation (Smialowicz et al., 1984c). The clearance of radiolabeled tumor cells from the lungs of Mn-treated mice was also significantly enhanced compared with saline controls. As might be expected, Mn-treated mice had significantly fewer lung tumor foci compared with saline controls following the injection of B16-F10 melanoma cells (Table 1).

When Ni and Mn were injected concomittantly in contralateral thighs, the suppression of NK activity by $NiCl_2$ was preempted by $MnCl_2$ (Table 1). Injection of $MnCl_2$ one day prior but not one day subsequent to $NiCl_2$ also resulted in enhanced NK cell activity (Rogers et al., 1983).

DISCUSSION AND CONCLUSIONS

In addition to suppressing NK cell activity, $NiCl_2$ suppresses thymus-derived or T lymphocyte responses in the mouse (Smialowicz et al., 1984a,b). Both of these lymphocyte populations play a role in immune defenses against neoplasia. In contrast, $MnCl_2$ enhances NK cell activity as well as macrophage function, including macrophage tumoricidal activity (Rogers, et al., 1983; Smialowicz et al., 1984c,d). Mn-enhanced natural cell-mediated immune effector cell function appears to be mediated by interferon. These results taken together may provide some insights into the inhibitory effect of Mn on Ni-induced sarcomas in rats described by Sunderman and McCulley (1983). While the results are far from conclusive, they suggest that an epigenetic mechanism may contribute to Ni-induced carcinogenesis.

ACKNOWLEDGMENTS

The author acknowledges R.R. Rogers, M.M. Riddle, R.W. Luebke, D.G. Rowe, and R.J. Garner for their contributions to this work and to F. Gregory for typing this manuscript.

DISCLAIMER

This paper has been reviewed by the Health Effects Research Laboratory, U.S. Environmental Protection Agency, and approved for publication. Mention of trade names or commercial products does not constitute endorsement or recommendation for use.

REFERENCES

Adkins B., Richards J.H. and Gardner D.E. (1979) Environmental Research, 29, 33-42.

Herberman R.B. and Ortaldo J.R. (1981) Science, 214, 24-30.

Rogers R.R., Garner R.J., Riddle M.M., Luebke R.W. and Smialowicz R.J. (1983) Toxicology Applied Pharmacology, 70, 7-17.

Smialowicz R.J., Rogers R.R., Riddle M.M. and Stott G.A. (1984a) Environmental Research, 33, 413-427.

Smialowicz R.J., Rogers R.R., Riddle M.M., Garner R.J., Rowe, D.G. and Luebke R.W. (1984b) Environmental Research (in press).

Smialowicz R.J., Rogers, R.R., Riddle M.M., Luebke R.W., Rowe, D.G. and Garner R.J. (1984c) Journal of Immunopharmacology, 6, 1-23.

Smialowicz R.J., Luebke R.W., Rogers R.R., Riddle M.M. and Rowe D.G. (1984d) Immunopharmacology (Submitted).

Sunderman F.W., Jr. (1981) Environmental Health Perspectives, 40, 131-141.

Sunderman F.W., Jr. and McCully K.S. (1983) Carcinogenesis, 4, 461-465.

461640

NICKEL INDUCED OCCUPATIONAL ASTHMA : IMMUNOLOGICAL AND CLINICAL ASPECTS

ANGELO M. CIRLA*, FERRUCCIO BERNABEO*, FABRIZIO OTTOBONI** RENATA RATTI*

* Occupational Health Unit, Hospital of Lecco, 22053 Lecco, Italy

** Lofarma Research Laboratory, viale Cassala 40 20143 Milano, Italy

INTRODUCTION

Reports of asthmatic lung disease in nickel-exposed workers have been fairly rare till now. Only 4 single patients were separately studied by Talot et al.(1957), McConnel et al. (1973), Malo et al.(1982) and Novey et al.(1983). All these sporadic cases occurred typically in electroplating industry, where an exposure to nickel sulfate or nickel chloride aerosols is common, usually at low airborne concentrations.

Our investigations concern 12 workers in nickel plating operations; their complaints were specifically referred to their physical activity, consisting of recurring worsening respiratory distress.

SUBJECTS AND CLINICAL HISTORY

Four women and eight men were admitted to our Unit during the last 5 years. They worked in small metal plating factories. In one case respiratory troubles were unfounded and an hand dermatitis was the only illness. 7 subjects were typically asthmatics, with a clinical picture of wheezing and bronchial obstruction ascertained in the past or at present, during the physical and functional examination; they had experience of respiratory distress on the working-days, were forced to stop activity, improved during the rest period and relapsed a few days after the resumption of their jobs. Cough, wheezing and dyspnea were generally reported as more severe after the work-shift and at night. There was no associated fever, chills, chest pains or dark sputum; in 3 cases hand dermatitis was also present. The other 4 workers had prevailing symptoms of rhinitis and chest tightness, but they did not ever experience true asthmatic attacks.

Clinical examination

All subjects underwent the following baseline tests: blood cell count, ESR, chest X-ray, pulmonary function tests: specific bronchial provocation test was also performed (mist of distilled water method). According to the findings, the diagnosis of asthma, rhinitis and dermatitis was confirmed.

Immunological studies

A skin test panel of 15 common allergens was adopted by intradermal technique, in order to evaluate the atopic status.
Patch-tests (PT) were applied to nickel sulfate, nickel chloride, potassium dichromate, copper sulfate, cobalt sulfate, with evaluations after 20 minutes and 48 hours.
Immunoglobulins classes (immunodiffusion test) and total serum IgE (radioimmunosorbent assay) were determined.
Nickel-specific IgE antibodies were determined according to the method suggested by Novey et al.(1983), modified. Our technique adopts a particular conjugate made of a cationic exchange resin (Dowex 50W X8 - B.D.H.Chemicals) saturated with nickel sulfate ($NiSO_4.6\ H_2O$), as insoluble matrix in the RAST.
Nickel-specific IgG antibodies were tentatively investigated by a similar radioimmunoassay, using labelled serum protein A.

Challenge studies

Bronchial provocation tests (BPT) were performed in an exposure chamber, where low concentrations of salt solutions may be nebulized to obtain definite airborne concentrations. For nickel sulfate aerosols (according to ACGIH STEL-limits) a 0,3 mg/m^3 concentration was routinely selected and the exposure time was fixed at 30 min.
Functional monitoring (FEV_1) was realized before and after the exposure during 24 hours. A physician was in attendance throughout the whole period, to perform clinical observation and to treat any severe reaction if necessary. A $>$15% drop in FEV_1, compared with the control basic value, was accepted as positive response.

Control challenges

Some subjects were also exposed to copper or chromium salts aerosols in dependence of their occupational history. Some others,with a job only in nickel plating, were challenged by the same BPT using iron sulfate as control substance. Moreover two atopic asthmatic volunteers were submitted to the nickel challenge at the same airbone concentrations of 0,3 mg/m^3.

RESULTS AND DISCUSSION

A significant broncoconstriction caused by nickel inhalation was registered in 6 workers,while all others were negative. Since the bronchial reactivity was unaffected by challenges to other metal salts, the positive responses seemed to be quite specific. As shown in table I, all responders were asthmatic and no correlation was observed with other clinical pictures.

TABLE I. Nickel hypersensitivity in 12 patients

Syndrome	Subjects (number)	PT+	BPT+
Rhinitis and Chest tightness	4	0	0
Asthma and dermatitis	3	3	3
Asthma	4	0	3
Dermatitis	1	1	0

The BPT responses were of dual or late type;the late-onset reactions were more severe and protracted in all cases. Nickel-specific IgE (table 2) were detected in the sera of 3 patients, who showed a dual (immediate and late) reaction after the exposure. The presence of Nickel-specific IgG was equivocal by the assay method adopted and the reliability of the results is open to question.

In all cases no correlation was observed with atopic status, bronchial aspecific hyperresponsiveness,smoking habits and occupational exposure period. Nickel patch-tests were positive only if dermatitis was associated.

TABLE 2. Immunological findings in 6 nickel asthmatics

Subject	Age (years)	Exposure (period)	Atopy	Ni-IgE	Ni-BPT
M.M.	47	3 yr	not	pos.	dual
G.R.	40	2 yr	not	pos.	dual
D.L.	42	5 yr	not	pos.	dual
C.S.	35	I yr	yes	neg.	late
B.M.	54	5 yr	not	neg.	late
E.P.	49	3 mo	not	neg.	late

CONCLUSIONS

The study confirm the possibility of selective respiratory sensitization by nickel fumes or aerosols. Nickel is a well known hapten and its combination with proteins, mainly serum albumin,might develop antigenic determinants in the bronchial tree. A part of response, at least,was demonstrated to be mediated by an IgE immunologic mechanism.
It is probable that unusual episodic pollutions in the work environment,like galvanic insufflation process, could stimulate the immunological reactivity in predisposed workers, giving rise to the allergic asthma.
The bronchial challenge is at present the most reliable method for the diagnosis of nickel-asthma.

REFERENCES

Malo J.L.,Cartier A.,Doepner M.,Nieboer E. (1982) J. All. Clin.immunol. 69,55-59

McConnel L.H.,Fink J.N.,Schlueter D.P.,Schmidt M.G. (1973) Ann.Int.Med. 78, 888-890

Novey H.S.,Habib M.,Wells J. (1983) J. All. Clin. Immunol. 72,407-412

Talot F.,Brondeur P.,Neulat G. (1957) Arch. Malad Profess. 18,291-295

BIOLOGICAL CONSEQUENCES OF LONG-TERM EXPOSURE TO ORTHOPEDIC IMPLANTS

H.F.HILDEBRAND[1], B.ROUMAZEILLE[2], J.DECOULX[2], M.C.HERLANT-PEERS[3], P.OSTAPCZUK[4], M.STOEPPLER[4], and J.M.MERCIER[1].
1. Institut de Médecine du Travail 2. Service de Traumatologie and 3. U.124 INSERM, 1 Place de Verdun, 59045 LILLE-Cédex, France
4. Institut f. Angewandte Physikalische Chemie Kernforschungsanlage, Jülich, GFR

INTRODUCTION

Co-Cr-Ni alloys are frequently used for orthopaedic implants in spite of the increasing number of reports dealing with reactions of intolerance to these metals. In addition clinical and scientific research suggest a relation between the use of these alloys and the increase of allergic reactions to the metals contained.

In the present work, the metabolism of the diffusing metal ions is investigated in 10 patients after removal of orthopaedic implants : metal-on-metal total hip implants, osteosynthesis plates and centro-medullary nails.

MATERIAL AND METHODS.

Following biological materials were collected from patients : blood, urine, hair, finger- and toenails and implant surrounding tissue samples.

Ultrastructural studies were carried out as described previously (Hildebrand and Biserte, 1978). EDS microprobe analyses were performed on unstained sections placed on Pioloform-F coated titanium or gold grids. Elemental determinations were carried out with a Philips EM 301 electron microscope and a Link-System microprobe analyser.

AAS analyses of Ni and Cr were performed on a Perkin-Elmer 5000 Spectrometer (Stoeppler, 1980; Sunderman, 1980; Veillon et al. 1982). Additional Ni and Co analyses were carried out by voltametric determination (Ostapczuk et al. 1983).

RESULTS AND DISCUSSION

Some morphological changes of the tissue surrounding implant can be observed. Connective tissue displays an increase of cell density and high quantties of metallic intra- and extracellular precipitates. The intracellular deposits are generally observed in macrophages, fibroblasts and multinuclear giant cells. Ultrastructural examination reveals quite different aspects of these structures (Fig.1), varying from the patient, the implant type and the examined tissue.

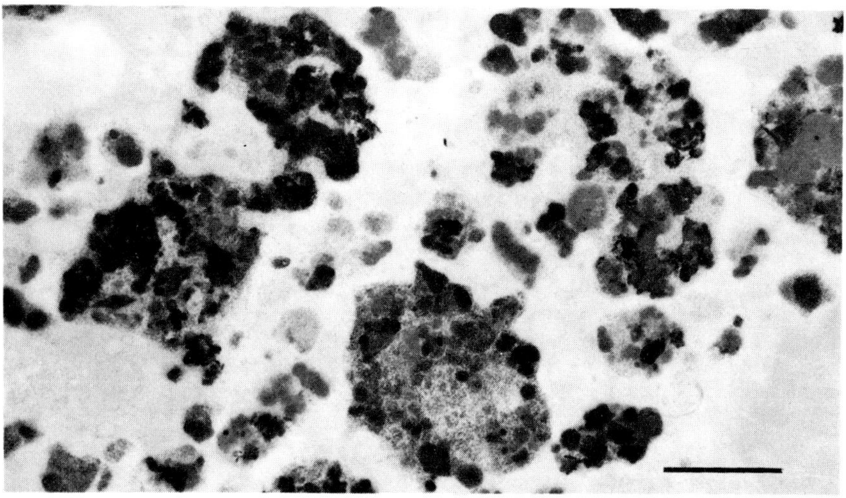

Fig.1. Precipitate in a multinuclear giant cell from connective tissue in the vicinity of a hip prosthesis. Unstained section. Bar:1 μm.

The particles and precipitates were analysed by X-ray energy dispersive spectrometry (EDS). The elementary composition of the particles and precipitates varies as a function of the morphological differences and, of course, of the initial composition of the implant alloy.

In tissue samples next to osteosynthesis plates, Cr, Fe and Ni are found to be the main components of crystalline structures. In more organic structures, Cl, Ca, S and especially P are also present (Fig.2). The presence of Fe is certainly due to haemoglobin-Fe deriving from haemorrhage after placement of the orthopaedic implants. The rather high quantity of P may be extracted from bone. Its association to Cr suggests the existance of a Cr-P

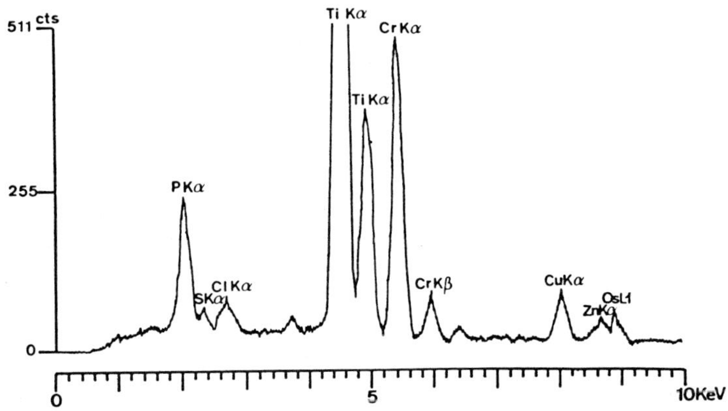

Fig.2. EDS-analysis of organic deposits in tissue adjacent to osteosynthesis plate. Note the presence of S, Cl and P peaks.

complex which could be considered as a Cr-metabolite.

Co is never detected in the precipitates but is generally revealed in blood, plasma and urine, sometimes even in a more than ten-fold concentration (Table 1). A clear increase of Ni and Co concentration is evident after long term exposure. Ni and Cr concentrations are especially high in biological fluids from patients with complications (Hildebrand et al., 1982; Dobbs and Minski, 1980).

High concentrations of Ni and Co could also be detected in tissue samples. The high Co concentrations are surprising, since we never revealed Co in the precipitates by EDS analyses. Thus Co may be bound to whole tissue in an unprecipitated form not detectable by the analysis of a very small aera.

CONCLUSIONS

Orthopaedic implants of Co-Cr-Ni alloys can corrode in biological milieu and the corrosion product is recovered in biological fluids as well as in implant surrounding tissue.

The presence of other elements (S, Cl, Ca and especially P) suggests the metabolization of the corrosion products and the existence of metallo-organic metabolites.

In constrast to occupational exposure orthopaedic implants are an internal- quite distinct - source of allergenic and toxic elements. Sometimes alarming high concentrations of these elements are recovered in tissues and body fluids, so that more intensive surveillance of patients with such implants is to be recommended.

implants	exp time(y)	Plasma ng/g			Blood ng/g			Urine µg/g creatinine		
		Ni	Co	Cr	Ni	Co	Cr	Ni	Co	Cr
total hip replacement	12	5.00	1.54	8.43	2.59	0.33	27.08	6.3	2.63	112.04
osteosynth plate	4	1.77	0.29	0.65	1.76	0.21	0.86	3.33	0.26	0.47
osteosynth plate	8	7.92	0.63	1.38	4.60	0.44	2.41	—	—	—
osteosynth plate-nail with complications	8	23.7	0.27	3.67	11.41	0.15	—	0.75	0.18	1 29
nail	2	2.30	0.42	—	1.25	0.33	—	3.90	0.25	—
nail	8	7.92	0.63	—	4.60	0.44	—	—	—	—
normal values		<0.9	<0.10	<1.40	<1.80	<0.11	<3.00	<3.3	<0.36	<3.00

Table. 1. Ni-Co-Cr concentrations in biological fluids of patients with orthopaedic implants.

REFERENCES

Dobbs H.S. and Minski M.J. (1980) Biomaterials,1,193-198.

Hildebrand H.F. and Biserte G. (1978) Cancer, 42,528-554.

Hildebrand H.F., Herlant-Peers M.C., Stoeppler M., Bagschik U. and Roggatz J. (1982) in Hagenoer J.M. and Furon D. (Eds) Actes des XVIIe Journ. Nat. Med. Travail, pp 124-126.

Ostapczuk P., Stoeppler, M. and Nürnberg H.W. (1983) in Brown S.S. and Savory J. (Eds) Proc. 2nd Int. Conf. Clin. Chem. & Chem. Toxicol. Metals pp 61-64. Academic Press, London.

Stoeppler M. (1980) in Nriagu J.O. (Ed.) Nickel in the Environment pp 661-821. J. Wiley, N.Y.

Sunderman F.W. Jr (1980) Pure Appl.Chem.52,527-544.

Veillon C.,Patterson K.Y. and Bryden N.A. (1982) Clinical Chemistry 28, 2309-2311.

461641

CYTOLOGY AS A SCREENING METHOD IN THE DETECTION OF PRECANCEROUS NASAL MUCOSAL CHANGES IN NICKEL WORKERS.

RAGNAR VOSS, ALBRECHT REITH AND
SISSEL REICHBORN-KJENNERUD

Department of Pathology, The Norwegian Radium Hospital and Norsk Hydro's Institute for Cancer Research, Montebello, Oslo 3, Norway and The Norwegian Cancer Society.

INTRODUCTION

The carcinogenicity of nickel and nickel compounds is well known from in vivo and in vitro studies (Reith and Brøgger, 1984). A higher incidence of carcinoma of the lung, the nasal cavities and the sinuses in nickel workers was found in epidemiological studies by Pedersen et al. (1973) and Magnus et al. (1982). In a clinical and histological study of these nickel workers in 1979, Torjussen et al. found a high frequency of metaplastic and dysplastic changes of the nasal mucosa, confirmed in 1982 by Boysen et al. They proposed a concept of gradual transformation of the normal respiratory epithelium to metaplastic and possibly to dysplastic epithelium. Morphometric analysis demonstrated a stepwise transition from the ciliated, pseudostratified columnar epithelium to stratified cuboidal, stratified squamous and dysplastic epithelium (Reith et al. 1983). The latter is regarded as a precancerous lesion. However, only a certain proportion of dysplastic epithelium develops into carcinoma. Since no figures have been estimated for the nasal epithelium, the question of which dysplasias turn into cancer is still unanswered. Sinonasal carcinoma is an often fatal disease, and diagnosis often delayed until its extension has provoked symptoms which lead the patient to the doctor. Early diagnosis would improve chances of survival. Screening of selected risk groups is therefore important for detection of tumours before their growth has made curability impossible.

Earlier investigations of the nasal mucosa of nickel workers have been based on the resource-consuming biopsy procedure. In many medical specialities exfoliative cytology is well-established as a simple and representative method for mass screening for precancer and cancer. The aim of the present study was to find out if this non-invasive technique would yield suitable material for light microscopical (LM) and scanning electron microscopical (SEM) examination of exfoliated nasal epithelial cells.

MATERIAL AND METHODS

A group of 36 workers with more than 10 years employment in a nickel refinery was examined. The cellular material was obtained by a small brush. The anterior part of the middle nasal turbinate was gently brushed after superficial anesthesia with Tetracain 1.5%/Naphazoline 0.25%, which also decongested the mucosa, making the access better (Fig. 1). A small part of the anterior curvature of the middle turbinate was left unbrushed. from where the biopsy was simultaneously obtained. Histological examination served as a control.

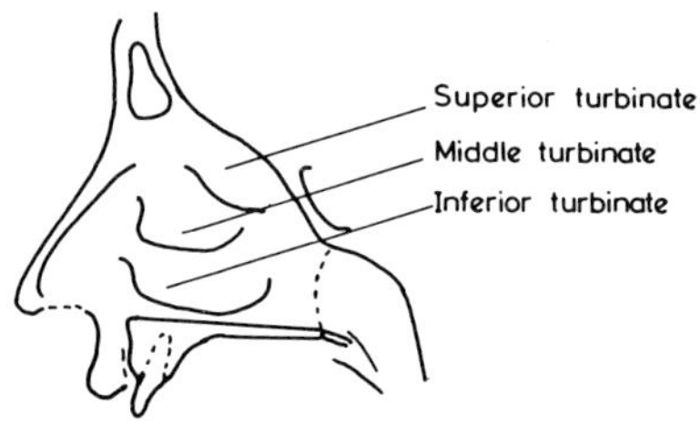

Fig. 1. Lateral wall of nasal cavity.

After immediate fixation. the smears were stained according to the Papanicolaou procedure. The histological sections were stained with hematoxylin and eosin. The specimens were examined by a cytopathologist and a histopathologist respectively and independently. The results were not compared until final diagnoses were made.

The SEM specimens were obtained from the same brush wiped off on to a cover glass and fixed. These smears were later dehydrated, critical point dried and coated with gold.

RESULTS AND DISCUSSION

The method yielded excellent material for cytological examination. Ciliated columnar cells from the normal respiratory epithelium (Fig. 2). metaplastic cells as well as dysplastic epithelial cells (Fig. 3) could be recognized.

The cytological and histological diagnoses corresponded in 26 of the cases (72%).

The smears for SEM examination were excellent for discrimination of subcellular structures on the luminal and intercellular surfaces of the cells (Fig. 4), clearly displaying the diagnostically important microvilli pattern.

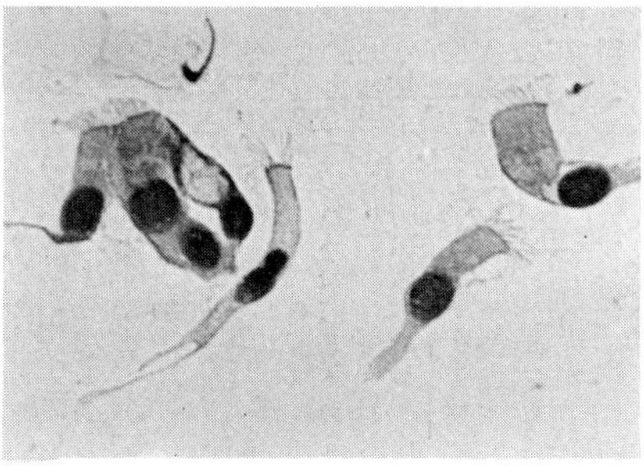

Fig. 2. Ciliated, columnar cells from normal respiratory epithelium (X 440).

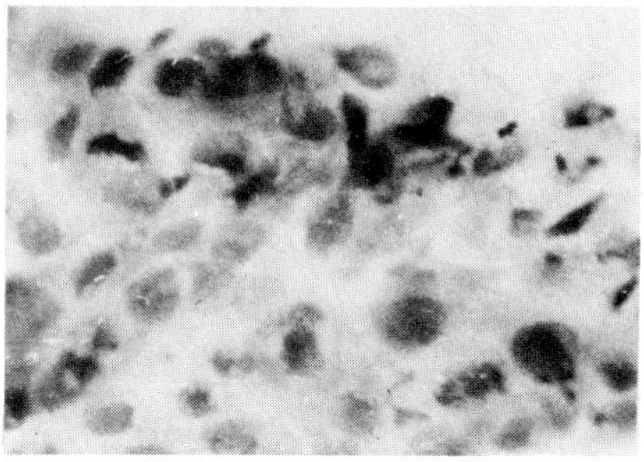

Fig. 3. Dysplastic cells with great variation in nuclear size and density (X 440).

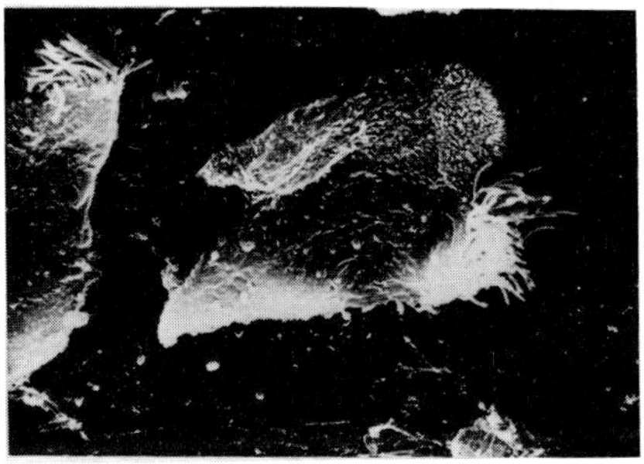

Fig. 4. Ciliated cells and a goblet cell with microvilli on the luminal surface (X 1300).

Reith et al. (1983) have shown that normal, metaplastic and dysplastic epithelium can be discriminated by morphometric analysis in histological sections. The cellular material obtained in the present study was found suitable for morphometry of the nuclei. Further studies are in progress to find out if such analysis can be applied to cytological smears.

A review of the available literature on nasal cytology gave only sparse information about epithelial cells and nothing about the diagnosis of dysplasia. Most previous publications described the non-epithelial cells in infectious and allergic disease.

Cytological screening has many advantages compared to histological examination: The cytological smear can be obtained more quickly than a biopsy. Even without anesthesia the discomfort is minimal and brief. Only gentle brushing is necessary and thus bleeding can be avoided. The workers can therefore return directly afterwards to the production. Biopsy taking leaves a small wound and causes slight bleeding, and the workers usually stay out of production for 1-2 days. Consequently, cytological sampling is only of negligible disturbance to industrial production compared to biopsy screening.

Cytological specimens are simpler to process than histological sections. Microscopical screening of the smears is, however, more time consuming and more difficult than histological examination.

The anterior part of the middle nasal turbinate is the most susceptible area for deposition of airborne particles (Hadfield. 1970). In earlier examinations (Torjussen et al., 1979. Boysen et al.. 1982), the biopsies were obtained from this area. Exfoliative cytology by brushing collects cells from a larger and therefore more representative area than a restricted biopsy. This is important, since a dysplastic focus may vary from a fraction of to more than a millimeter in size. and the changes are also multifocal (Reith et al.. 1983).

CONCLUSION

Exfoliative cytology of the nasal mucosa seems to be a simple and reliable method for routine examination of risk groups. The preliminary results reported here encourage further evaluation of the technique.

REFERENCES

Boysen. M.. Solberg L.A.. Andersen I.. Høgetveit A.C. and Torjussen. W. (1982) Scandinavian Journal of Work. Environment and Health 8. 283-289.

Hadfield. E. H. (1970). Annals of the Royal College of Surgeons of England 46. 302-319.

Magnus. K.. Andersen A and Høgetveit A.C. (1982). International Journal of Cancer 30. 681-685.

Pedersen. E.. Høgetveit A.C. and Andersen A. (1973). International Journal of Cancer 12. 32-41.

Reith. A. and Brøgger A. (1984). in Nickel in the Human Environment. Proceedings of the symposium on Nickel in the Human Environment. Lyon. 8-11 March 1983. IARC (in press).

Reith. A.. Rigaut J-P.. Boysen. M. and Marton. P. (1983). in Reznick G. and Stinson S.F. (Eds.). Nasal Tumors in Animals and Man. vol. II. pp. 55-94. CRC Press Inc., Boca Raton, Florida.

Torjussen W.. Solberg. L.A. and Høgetveit A.C. (1979). British Journal of Cancer 40, 568-580.

SUMMARY REMARKS OF SESSION CHAIRMEN (Stanley S. Brown, Mirna Mikac-Devic, John Savory, Markus Stoeppler, Jean Meininger and Jan Sjoholm)

Useful improvements in the analytical chemistry of nickel were described by several authors. Thus Dr. Christensen reported a rapid Zeeman atomic absorption procedure for determining nickel and cadmium in heparinized blood. Results for determinations of these elements in blood collected in 1983 from the general population of Denmark were given. The simplicity of the procedure lessens the risk of contamination and makes the method attractive as a screening technique.

Dr. Ostapczuk used the new EG&G 384B Polarographic Analyzer for the determination of nickel and cobalt by square wave voltametry at the hanging drop electrode in alkaline solution. Concentrations of 1.5 and 0.15 ng/ml were found for nickel and cobalt, respectively, in whole blood of normal subjects. In persons exposed to stainless steel implants, the values were an order of magnitude higher.

Dr. Nieboer showed that furil-α-dioxime (FD), like dimethylglyoxime, sensitized the differential pulse polarographic determination of nickel and cobalt. Since currents of 5 nA can now be measured accurately and precisely, FD permits a detection limit for Ni^{2+} of 0.02 µg/l, and for Co^{2+} of < 0.01 µg/l, providing adventitious contamination of buffers and reagents is avoided. Because interference from other metals is minimal, the technique is eminently suitable for the simultaneous determination of ultratrace quantities of these elements.

Different groups of papers described some applications of quantitative measurements of nickel in clinical and occupational medicine. Dr. Hopfer measured nickel concentrations in serum and whole blood from healthy control subjects, patients with end-stage renal disease who had not been treated by hemodialysis, and patients with end-stage renal disease undergoing hemodialysis. Hypernickelemia was found in all hemodialysis patients and the authors speculated that this contributes to hemodialysis-associated hypersensitivity, encephalopathy, and arteriosclerosis.

Dr. Savory investigated both serum and lymphocyte aluminium and nickel concentrations in normal subjects and patients on hemodialysis. The results demonstrated significant elevations of serum Al and Ni and of lymphocyte Al concentrations in patients on hemodialysis. Thus, nickel may contribute to the complications observed in these patients, which so far have been mainly attributed to aluminium toxicity.

Dr. Drazniowsky reported an improved method for serum nickel determination, using flameless atomic absorption, in which elaborate pre-treatment/extraction procedures were avoided. The method was used to determine serum nickel in normal subjects and patients with chronic renal failure. Complete matrix matching of samples and standards was obtained by the use of serum-based standards, which gave parallel calibration curves to those for normal and uraemic serum samples. In haemodialysis patients, pre-dialysis serum nickel concentrations were significantly higher than in normal subjects.

Dr. Leach presented the results of carefully conducted controlled studies on serum nickel concentrations in angina and myocardial infarction. Earlier observations pointing to high concentrations in myocardial diseases were confirmed. Moreover, the values found for serum nickel in healthy adults were the lowest so far reported, which testifies to the accuracy of the analytical method and the validity of the sampling strategy. However, the author's observations about some fluctuations in results for individuals at these extremely low levels may be due to experimental artefacts rather than to true physiological variations.

The papers by Dr. Aitio and Dr. Schaller pointed up the practical problems of monitoring occupational exposures to nickel and chromium in their various forms. Because the two investigations were performed with quite different external exposure conditions, a direct comparison is not possible between the results obtained, e.g. in respect of estimations of biological half-lives of nickel in urine, and the significance of nickel concentrations in urine during the work shift, after shift etc. Further studies need to be carried out over longer periods to confirm these findings. From both papers, however, it is evident that urine collected in a clean, nickel-free environment can be useful for workplace-control purposes. Both from the sampling and analytical viewpoints, it is clear that urine offers distinct advantages over serum for routine monitoring.

Several interesting studies on different aspects of the pathology of nickel were presented in these sessions. Thus Dr. Smialowicz discussed the effects of nickel and manganese chlorides on natural killer (NK) cell activity in rodents. NK cells are lymphocytes which are believed to be one of the first lines of non-specific defence of the immune system. Nickel chloride caused a significant reduction in NK cell activity, and this enhancement was associated with increased circulating interferon. These results may provide important clues to understanding the carcinogenic potential of nickel compounds and the antagonism of manganese previously described by F. W. Sunderman Jr. and his colleagues.

Dr. Cirla reported the results of clinical examination of "hypersensitive" workers in the electroplating industry - 7 cases of asthma, 4 of rhinitis and 1 with thoracic pains, together with 2 volunteers. The diagnostic procedures involved clinical examination, skin tests and bronchial provocation challenges, to which six cases were typically reactive. Immunological studies were performed, and the occurrence of anti-nickel IgE was demonstrated in some cases. These findings confirm the possibility of respiratory sensitization by nickel vapours or fumes, but environmental control and medical surveillance should serve to avoid these occupational diseases.

Dr. Hildebrand studied the metabolism of diffusing metal ions in 10 patients with metal-on-metal total hip implants, osteosynthesis plates and centro-medullary nails. The ultrastructural studies revealed metallic intra- and extra-cellular precipitates in the tissues around the implants. Chromium and nickel were found to be the main components of these bodies. Cobalt was never detected, but it was found in body fluids, in which Ni and Cr concentrations are specially high in patients with complications. It is clear that orthopaedic implants may be an internal source of allergenic elements.

Dr. Voss showed the value of a cytologic method for obtaining cell samples of mucosa by brushing. With this technique, samples can be taken from a wide area of the mucosa to yield excellent preparations, very suitable for diagnosis by means of light microscopy and morphometric analysis. Precancerous dysplasia may be detected with this cytological technique, which has obvious advantages over conventional biopsy.

The immunological and clinical aspects of these four papers are of interest both to researchers and to medical practitioners. The transitory effects of nickel and manganese on the cells of the immune system warrant further study. For example, the duration of the presence of nickel and manganese could be elucidated, as well as the possible role played by manganese in activating natural killer cell enzymes. Such data might open up new lines of research in carcinogenesis.

The observations on orthopaedic implants justify medical surveillance of such patients. It is generally recommended that osteosynthesis plates be removed whenever possible, but the metabolism of the diffusing metal ions must involve several different factors. Thus research performed in a British hospital on 13 patients with total hip implants did not reveal any significant amounts of nickel in the body fluids and tissues.

All those concerned with health surveillance of nickel workers found the cytological method for obtaining cell samples of nasal mucosa by brushing to be of great interest, since it requires no more than a contact anesthetic. Although the discovery of cellular dysplasia provides a further argument in favour of closer medical surveillance, it does not justify removal of the worker from the place of employment.

OCCUPATIONAL EXPOSURES

461642

ATMOSPHERIC MONITORING OF NICKEL-CONTAINING DUSTS AT THE INCO REFINERY IN CLYDACH, WALES

LINDSAY G. MORGAN.

Medical Department, INCO Europe, Clydach, Wales

INTRODUCTION

The main purpose of the current interest in the toxicology of nickel and its compounds is to assess their relative dangers to man and to establish levels of exposure which do not produce adverse health effects. Animal studes, almost all of them using routes other than respiratory, support the view that some forms of nickel produce local tumors upon injection. Epidemiological data however, suggest that certain operations in nickel refining have been associated with increased risk of nasal and lung cancer. These same epidemiological studies however, also suggest that men can, and have worked with exposure to airborne dust containing nickel and nickel compound without experiencing any increased risk of developing respiratory cancer.

It is still not known with any degree of certainty which nickel compounds are carcinogenic to man or under what circumstances. In the absence of certain knowledge, some governmental agencies have decided upon, or are currently considering, what levels of exposure are acceptable or reasonable and whether individual standards should be set for different nickel species. Once the appropriate decisions have been reached then it will be the turn of the Occupational Hygienist to ensure that measurements of the atmosphere are as accurate as possible and to define the limits of accuracy so that results are meaningful and so that the inspector can ensure compliance. The subject of atmospheric sampling has been exhaustively reviewed by WHO (1984), and NiPERA (1984) has prepared a document on sampling of airborne nickel. [The purpose of this paper is to outline the procedure used at the Clydach refinery to provide a basis for discussion and to stimulate the production of a standard procedure for the measurement of nickel and its various species in the working atmosphere, just has been done for the measurement of nickel in urine and blood by the Nickel Subcommittee of IUPAC.] (1)

CLYDACH REFINERY PROCEDURES

Details of the refinery process have been published elsewhere (Morgan1958; Morgan 1979). The first record of atmospheric monitoring dates to 1932. A sample of 100 litres of air was pulled through a cotton wool plug which was subsequently analysed for nickel content. In 1932 the plant was not running at full capacity and the results ranged from 8-42 mg Ni/m^3 so in all probability higher levels were the norm during full production, particularly in the old calciners department.

No more monitoring appears to have been carried out until 1945 when particle dust counting was used. This cannot be related to gravimetric analysis but it is interesting to note that in the 1945 and 1949 surveys, figures from 1.7 to 19.1 million particles per cubic foot (mppcf) less than 10 μm, were recorded. By comparison, the current TLV for Quartz is 3 mppcf and for nuisance dusts 30 mppcf.

The current programme at Clydach aims to monitor all employees who may be exposed to nickel containing dusts for the whole of one working shift, twice a year. The programme is voluntary, only about 5% of personnel refuse to cooperate. The pumps are standard Casella units with a flow rate of 2 litres/min. This is checked at the beginning and end of the shift. The filter is a GF/A with a pore size of 0.5 μm and the sampling head is modified to avoid fallout and prevent the face turning against the clothing. Since the TLV does not specify the particle size and since one of the risks is nasal cancer, no cyclone is fitted. Analysis is carried out using X-Ray spectography. Where estimation of soluble nickel is required it is performed by solution of the filter in water and then analysis by atomic absorption spectrophotometry under standard conditions. Static sampling is used for special studies of pieces of equipment or where information is required about dust levels at specific locations but it is not used as a routine procedure.

Employees are given results individually by letter or by interview with the Medical Officer or the plant manager. Group results are sent to managers and the departmental safety committee for discussion at that committee. Any high results are the subject of discussion between managers and the employee. Results are retained in perpetuity and are available both to the employee and employer.

It is important to consider the *sources of error*.

The pump is calibrated at the beginning and the end of the run. If the loss in flow is greater than 10% then the sample is discarded and the employee asked to wear the pump another time. Variations within the 10% limit are accepted and the mean flow calculated as the mid point between the two readings. In practice the loss in flow is in the order of 2%.

In respect of the *sampling procedure*, King (1980) has shown how the position of the sampling head on the body, the shape of the sampling head and the type of filter can all affect the ability of the head to take a sample that is representative of the air in the breathing zone. In order to reduce positional variation, the sampling head is placed on the collar of the overall by the nurse/hygienist at the beginning of the shift but there is no guarantee that it will stay there all day or if removed that it goes back to exactly to same place. The errors here are unquantifiable but an estimate has been obtained by asking a single operator to wear the pump over a succession of average shifts. The variation appeared to be in the order of 20%.

Since the shape of the filter head remains constant it is not a problem as a source of day to day variation on a routine monitoring programme but if mandatory limits are to be developed then the design characteristics of the head will have to be specified.

The type of *filter paper* that is used is also important. Some attract and others repel dusts depending upon their electrostatic properties and those of the material being handled. In any new situation this important factor must be given due consideration but it is not a source of day to day error in routine monitoring at the refinery where it is considered that the fibreglass filter paper is the most suitable.

The method of *analysis* used at Clydach is X-Ray spectography and has already been described (Morgan and Rouge 1979). It depends upon the dust being evenly distributed over the filter paper and if this is not achieved then the sample is discarded and a repeat requested. Repeatability is in the order of ±10μg in the 500-100μg range. In a recent inter laboratory comparison carried out by NiPERA (1984) different methods of analysis were compared.

It was shown that there is a potential for variation between laboratories of as much as 68% and this subject will require considerable attention.

Summation of all these errors suggests that with an observed result of 1 mg/m³ the true value probably lies between 0.6 and 1.4 mg/m³.

The action level for individual results is 0.5 mg/m³ (half the TLV). If this level is exceeded, enquires are made and a repeat requested. The group results for each assignment are discussed at the safey committee meeting. A stratified distribution of results is made available to the safety committee, with those of the previous year for comparison.

Statistical representation of results has been fully discussed by Leidel et al. (1977), any group trying to develop a standard technique will need to take this aspect of the work into full consideration.

Future developments will include the introduction of methods for estimating specific forms of nickel and NiPERA is currently investigating the practicability of this.

CONCLUSION

When national agencies have legislated on working thresholds of nickel in the atmosphere it will be necessary to measure compliance. There are a number of sources of error in the sampling and analytical procedures and it will be important for industry and the inspectorate to develop protocals which keep the errors to a minimum to ensure that results are comparable and meaningful.

REFERENCES

King E. (1980) Techniques for the measurement of cadmium in air. Report of Seminar "Occupational Exposure to Cadmium" *Cadmium Assoc., 34, Berkeley Square, London, UK.*

Leidel N.A., Busch K.A., Lynch J.R. (1977) Occupational exposure strategy manual *National Institute of Occupational Health & Safety, USA.*

Morgan J.G. (1958) Some observations on the incidence of respiratory cancer in nickel workers *BJIM 15. 224-234.*

Morgan L.G. (1979) Manufacturing Processes - Refining of Nickel *Occup. Med. 29.1. 33-35.*

Morgan L.G. and Rouge P.J.C. (1979) A study into the correlation between atmospheric and biological monitoring of nickel in nickel refinery workers *Ann. Occup. Hyg. 22. 311-317.*

NiPERA (1984) Blakely S.J., Statistical Analysis of Data from an Inter laboratory Test Programme for the Determination of Nickel in Bulk Materials *NiPERA Project 82-06.*

NiPERA (1984) Roy B.R., Air sampling method for inorganic Nickel *NiPERA Project 82-06.*

WHO (1984) Evaluation of exposure to airborne particles in the work environment *WHO Offset Publication No.80.*

THE DETERMINATION OF OCCUPATIONAL EXPOSURE TO AIRBORNE DUST CONTAINING NICKEL

RICHARD M WAGG
Health and Safety Executive
403-405 Edgware Road
London NW2 6LN UK

INTRODUCTION

Legislation and recommendations for workplace exposure limits for metal and other dusts, with few exceptions, imply that total particulate will be measured by personal samplers. The situation is complicated by the absence of an agreed standard instrument or sampling procedure, many organisations employing particular samplers for historical reasons. Difficulties then arise in comparing data, and in the development of meaningful workplace exposure limits for regulatory purposes. The move toward harmonisation of workplace health legislation among members of the European Communities requires a greater degree of standardisation of sampling methods and strategies for setting exposure limits, and also for measuring compliance against those standards, than has hitherto been accomplished. Much more attention is being given to dust measurement and this paper outlines some of the background and procedures currently being adopted by the Health and Safety Executive in the UK for measurement of nickel and other airborne particulate dusts.

SAMPLER CONSIDERATIONS

Total particulate 'total airborne dust' is only naturally defined if there is a particle size beyond which any particles are so rare they make a negligible contribution to the aerosol mass. This condition is rarely met in the work-place, particularly with the greater emphasis which is being given to personal sampling.

There have been two lines of approach to this problem for the purposes of standards. The first is to make a definition of 'total particulate' that reflects its biological effect. This approach has the advantage that measured concentrations are likely to correlate with disease; risks will be recognised earlier and control is more cost effective. This approach is analogous to the use of respirable dust definitions for those substances that are most dangerous when they deposit

FIG. 1 Collection efficiencies of personal samplers mounted on an aspirated human model

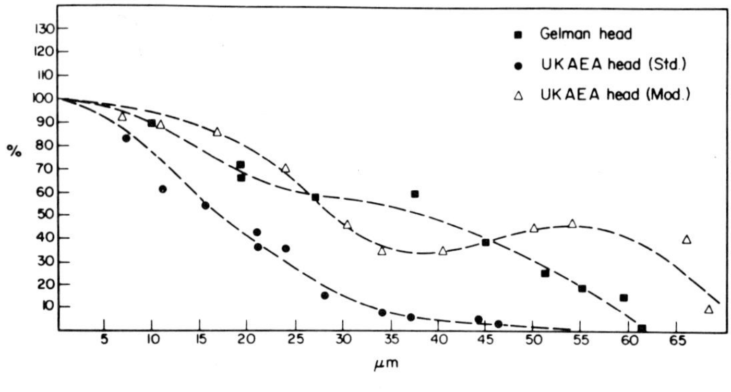

The open faced sampler exhibits a declining sampling efficiency curve with a cut off at about 60 μm. The single orifice head shows a much more pronounced declining characteristic with a cut off at about 50μm. The modified head exhibits characteristics similar to that of the open faced head but with a more complex characteristic sampling curve. The collection efficiencies of the modified head most closely approximate to the numerical values given in Table 1.

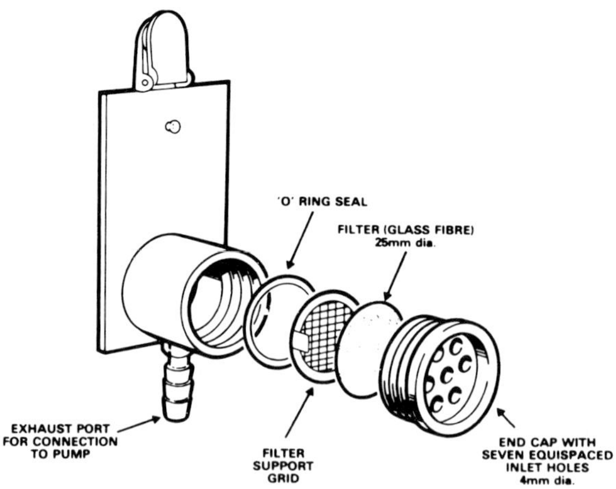

FIG. 2 Modified UKAEA personal sampling head used for total particulate sampling.

deep in the lung. The second approach is to take a particular sampler design and define in terms of what that sampler collects. This eliminates all the problems of designing a sampler to match a specified characteristic, but will only give good correlation with disease if the chosen design happens to match the biological effect.

An ad hoc Working Group under the ISO Committee TC 146, was set up to consider definitions of dust size ranges which may be used in preparing measuring methods for specific substances. The ISO(1983) report used the term 'inspirable' fraction, which was determined from directionally averaged entry characteristics of the human head, and includes particles that deposit immediately after entering the mouth or nose which may have a toxic effect if they are swallowed; a particle which does not enter will not present an inhalation or ingestion hazard.

TABLE 1 'Inspirable' fraction, as defined by the ad hoc Working Group of ISO Committee TC 146.

Particle aerodynamic diameter (μm)	0	4	12	20	50	100	185
'Inspirable' Fraction %	100	85.8	70.3	60.6	39.3	19.7	0

An alternative approach is to specify the sampler used for measurement without necessarily awaiting information on the size of particle collected. This procedure was adopted by the German Workplace Standards, MAK values, which defined 'total particulate' as that collected by a sampler with a entry velocity of 1.25 $m.sec^{-1}$. A personal sampler with a similar entry velocity and an entry orifice diameter of 4-10 mm has been adopted by the Council of European Communities Lead Directive (1982) for measurement of lead in the workplace. A similar direction has been followed by another working group of ISO TC 146 which has proposed to define 'total particulate' as that collected by a sampler with an entry velocity between 1.1 and 3.0 m. sec^{-1} at flow rates between 0.5 and 4.0 litres min^{-1}, corresponding to a range of circular diameter orifices of 2-9 mm.

Wood et al. (1978) have examined collection efficiencies of three types of personal samplers commonly used in the UK using an aspirated full scale human model in a wind tunnel. The three types all accept 25 mm filter discs and comprise an open face holder, the Casella Model T 13032 which is a single orifice closed face sampler, and a modified version of the same sampler with 6 additional 4 mm ports. The experimental results are shown in Fig. 1.

INSPIRABLE DUST SAMPLING

The measurement of workplace exposure using personal sampling techniques has the potential for significant error to occur. Use of a sampler which efficiently captures larger particles will lead to very pronounced enhancement of the measured concentration as larger particles form the highest proportion of the total mass collected. Closed face samplers are preferred for personal sampling and the Health and Safety Executive has adopted the modified head, illustrated in Fig.2, as the standard sampling instrument for the measurement of all particulate dusts in the workplace, excluding dusts containing lead.

Sample volume errors may also be a significant source of error in the final calculation of exposure concentration and procedures must be established to ensure these errors are minimised. Pumps should be fully charged and run for 15 mins. before calibration of the flow rate using an external calibrated flow meter, calibration should be rechecked at minimum intervals of two hours and always immediately before completion of the sampling.

Significant measurement errors will occur depending upon the location of the sampler in relation to the workers breathing zone, and its orientation. The sampling head should preferably be attached to a position on the worker's shoulder in such a way that its preferred orientation ensures that the filter is held close to the vertical position. Wherever possible the same sampler position should be consistently adopted for all measurements.

CONCLUDING REMARK

It is fundamental to the practice of occupational hygiene that an exposure limit has little real meaning except in the context of the measurement methods and strategies used. It is of primary importance to harmonise these latter factors first and then develop meaningful workplace exposure limits. The development of the multi orifice sampler and standardisation of sampling techniques is essential progress toward the establishment of relevant standards and means of assessing compliance with those standards.

REFERENCES

Council Directive 82/605/EEC

ISO Technical Report ISO/TR7708-1983(E)

Wood J.D. and Birkett J.L. (1978) Annals of Occupational Hygiene 22, 229-310.

461644

NiPERA 82-06 AIR SAMPLING METHOD FOR INORGANIC NICKEL

BERNARD R. ROY
Corporate Director of Loss Prevention, AMAX Inc.,
AMAX Center, Greenwich, Connecticut 06836

INTRODUCTION

Collection of personal or ambient samples of nickel metal and compounds (as Ni) entails drawing a known volume of air through a filter at a known flowrate. The filter consists of cellulose membrane, PVC or glass fiber depending on the preferences in analytical techniques. This paper describes an air sampling system and its calibration, and its application to total and respirable particulates.

SAMPLING SYSTEM

The sampler, or sampling system, consists of a pump, a flow control or indicating device (flowmeter or critical orifice), and a device for collecting the particulate. Equipment recommended includes:

1. Filter cassette: 2 or 3 piece containing a 37 mm filter. The filters that can be used are a a cellulose mixed-ester membrane (0.8 μm pore size), PVC or glass fiber filter. The analytical method, physical considerations (hygroscopy, static charge) and personal preference will determine the filter medium used.
2. Pump: A battery-operated vacuum pump with clip for attachment to the worker's belt, capable of operating at 2.0 liters per minute at a constant flow for at least eight hours.
3. Accessories: A length of tubing approximately two feet long to extend from the breathing zone "at clavical level" of the operator. For respirable samples a 10 millimeter nylone cyclone or other approved cyclone or cascade impactor capable of separating particles greater than 10 micrometers in diameter.

CALIBRATION OF EQUIPMENT

Sample pumps should be calibrated prior to and at the completion of each day's sampling, if possible, with a primary standard such as a soap bubble meter. Secondary standards (wet test meters or rotometers) may be used provided they are calibrated and adjustments are made for air pressure and temperature. To calibrate a personal sampling pump:

1. Check the voltage of the pump with a voltmeter to assure adequate voltage.
2. Place filter medium in holder and remove plugs.
3. The calibration train should consist of 2 liter bubble tube, cassette and sample pump.
4. Turn the pump on and moisten the inside of the soap bubble meter by immersing the buret in the soap solution.
5. Start a soap bubble up the buret and measure with a stopwatch the time it takes for the bubble to move from one calibration mark to another. (For a 2 liter bubble meter, time should be 1 minute.)
6. Repeat the procedure at least three times, average results, and calculate the flowrate by dividing the volume between the preselected marks by the time required for the soap bubble to traverse the distance.

SAMPLING PROCEDURE: (TOTAL PARTICULATE)

1. Remove the inlet and outlet plugs from the cassette. Total particulate sampling makes no distinction in particle size in that both respirable and non-respirable particulates are collected.
2. Adjust the pump flowrate for 2 liters per minute.
3. Attach the sample train on the individual to be sampled with the cassette at the "clavical level" as close to the breathing-zone as possible.
4. Turn the unit on, and record the time.
5. Check the unit when appropriate according to the reliability of the sampling equipment being used, the variability of the work environment, and job being performed, etc.
6. At the completion of sampling turn the pump off and record the time. Be sure to turn the cassette "face up" before turning off the pump to ensure any loose material on the filter medium does not fall off once the air flow has ceased.

7. The cassette should be sealed immediately after turning the pump off to prevent further contamination or loss of sample.

SAMPLING PROCEDURE: (RESPIRABLE PARTICULATE)

Respirable particulate sampling differentiates between those particle sizes which are small enough to enter the lung (generally considered to be less than 10 micrometers in size) and those that become trapped in the upper respiratory system. There are several means of isolating the respirable fraction but because of its availability and general acceptance for respirable particulate sampling, the 10 mm nylon cyclone will be discussed.

1. Calibrate the pump for 1.7 liters per minute (LPM).
2. Inspect the nylon cyclone to ensure that it is clean, dry and that scoring has not occured in the "cyclone" chamber.
3. Place the cassette into the cyclone holder with the entrance port facing down.
4. When placing the cylcone on the shoulder of the person being sampled, ensure that the inlet hole of the cyclone is facing outward and the cyclone is vertical. The higher pressure drop encountered using the cyclone may require that more frequent equipment inspections be performed to check pump flowrate.
5. Turn the unit on, and record the time and flowrate.
6. At the completion of sampling turn the pump off and record the time. Be sure to turn the cassette "face up" before turning off the pump to ensure any loose material on the filter medium does not fall off once the air flow has ceased.
7. The cassette should be sealed immediately after turning the pump off to prevent further contamination or loss of sample.

The cyclone separates respirable from non-respirable particles by causing an airstream of particles to change direction rapidly such that larger particles (over 10 micrometers in diameter) cannot make the change and fall into the lower chamber. The "respirable fraction" (less than 10 micrometers in diameter) is collected on the filter medium for analysis.

AMBIENT WORKROOM SAMPLING

It is generally well accepted that personnel sampling reflects more closely employee exposure than does sampling at fixed locations. On occasion, however, there is reason to conduct stationary sampling of the ambient workroom environment. Such reasons include identifying contaminant exposure potential, evaluating ventilation controls or providing more definitive sampling data to support personnel sampling results.

The total and respirable sampling methods described can be used for fixed location sampling. It is important to reinforce that these samples should be used to supplement a personnel sampling program, not as a substitute for it.

CONCENTRATION DU NICKEL ET DU CROME DANS L'ATMOSPHERE DES ATELIERS SIDERURGIQUES

F. KLEIN
Laboratoire d'Etude et de Contrôle de l'Environnement Sidérurgique, 57210 - Maizieres-les-Metz, France

La présente étude fournit un bilan des concentrations de chrome et de nickel relevées dan les prélèvements de poussières effectués à plusieurs postes de travail de différents ateliers sidérurgiques concernés : aciéries électrique et AOD, fonderie d'acier inoxydable, élaboration de fonte au chrome, ambiance de halle d'aciérie, atelier de chromage.

METHODOLOGIE

Sauf pour le cas des prélèvements aux postes de travail des ateliers de chromage qui ont nécessité le captage de l'aérosol liquide par barbotage, les poussières ambiantes ont été prélevées sur filtre en nitrate de cellulose à un débit d'aspiration de 40 m³/h.

Le nickel total et le nickel hydrosoluble ont été déterminés par spectrométrie d'absorption atomique. Le dosage du chrome total et du chrome VI hydrosoluble nécessite un protocole adapté. Celui-ci avait été défini préalablement pour le dosage des composés du chrome dans les poussières émises par les fours à arcs (Jecko, 1982) : (i) détermination du chrome total par spectrométrie d'absorption atomique après mise en solution acide (HCl + HNO_3) à chaud et reprise du résidu par fusion alcaline ; (ii) après extraction à l'eau, détermination du chrome VI par spectrophotométrie en présence de diphénylcarbazide pour éviter l'oxydation préalable de la solution ; (iii) utilisation de filtres en PVC pour éviter l'action réductrice des papiers-filtre.

RESULTATS

Le tableau I comporte les principaux résultats obtenus, présentés pour la plupart d'entre eux sous forme de moyennes de prélèvements:

	Ni total ($\mu g/m^3$)	Cr total ($\mu g/m^3$)
Four électrique		
- plancher	21	116
- halle de coulée	25	173
- salle de commande	2	15
Aciérie AOD		
- plancher	7	23
- halle de coulée	4	15
Halle d'aciérie		
- préparation des charges	20	40
- réfection poches	10	40
- meulage	100	300
- chalumage	141	37
Fondérie d'acier		
- épaillage	60	180
- décrochage	100	90
- ébarbage	80	60
- fonderie	60	50
Fonte au chrome		
- plancher	5	207
- salle de commande	1	31
Atelier de chromage		
- bains	-	18
- manutention	-	25

On constate que systématiquement : (i) la teneur en nickel hydrosoluble est inférieure à la limite de détection de l'analyse ; (ii) la concentration en chrome VI hydrosoluble reste toujours faible et pour tous les prélèvements effectués, inférieure à 2 $\mu g/m^3$ sauf pour les ateliers de chromage. Les résultats obtenus ici pour le chrome VI hydrosoluble ne correspondent pas a l'analyse du chrome dans le poussières émises par les fours à arcs (Jecko, 1982) dans lesquelles entre 0 et 33% pondéral du chrome total se retrouvait sous forme de chrome VI hydrosoluble. Cette distorsion n'a pas reçu d'explication à ce jour.

Les taux de nickel total et de chrome total ne dépassent jamais 10% du poids total des poussières en suspension prélevées. Nous avons noté que : (i) les taux de nickel et de chrome rencontrés ne dépendent pas de la composition de l'acier en cours d'élaboration ; (ii) les plus fortes teneurs en nickel proviennent du chalumage de lingots à forte teneur en nickel.

CONCLUSION

L'examen de la qualité de l'air ambiant aux postes de travail de plusieurs ateliers sidérurgiques vis-à-vis du nickel et du chrome montre que la teneur de ces deux éléments dans l'empoussiérage est toujours inférieure à 10% pondéral. De surcroît, on y note une absence de nickel hydrosoluble et une faible teneur systématique de chrome VI hydrosoluble. La toxicité des composés du nickel et du chrome dépendent largement des phases présentes et de l'état de valence. L'atteinte de ces paramètres est nécessaire à terme pour discuter de la toxicité des poussières mais également pour vérifier la méthodologie de prélèvement et d'analyse. Dans ce sens, des résultats prometteurs (Gandouin, 1984) ont été obtenus par microsonde à impact laser (LAMMA) pour la détermination de la valence VI du chrome dans les solides.

REFERENCES

Gandouin S. (1984) DEA, Université de Metz.

Jecko G. (1982) Rapport IRSID - ANA. RC 911.

UN NOUVEL APPAREIL DE PRELEVEMENT DE POUSSIERES, LE CIP 10

P. COURBON
Centre d'Etudes et Recherches des Charbonnages de France, B.P. No. 2, 60550 Verneuil-en-Halatte, France

INTRODUCTION

Les difficultés rencontrées dans la surveillance des empoussiérages miniers nous ont conduits, au Centre d'Etudes et Recherches de Charbonnages de France (CERCHAR), à étudier un nouvel appareil. Un de plus direz-vous, mais le problème posé par la mesure des empoussiérages en relation avec le risque effectif auquel sont exposés les ouvriers est un problème tellement important qu'aucun effort ne doit être négligé dans ce domaine.

Je ne connais pas ce qu'il en est dans l'industrie du nickel. Mais dans les mines de charbon où règne une endémie importante, les nombreuses études effectuées jusqu'à présent de par le monde montrent que le problème reste partout préoccupant.

Du point de vue technique, il est incontestable que la mesure individuelle à l'aide d'un appareil approprié porté par l'ouvrier au voisinage de ses voies respiratoires supérieures est la meilleure approche pour mesurer les quantités de poussières susceptibles d'être effectivement inhalées par les ouvriers au cours de leurs postes de travail.

Pour ce faire, l'appareil doit être:

de petites dimensions pour pouvoir être porté sans gêne par l'ouvrier;

effectuer un prélèvement à l'image des voies respiratoires de l'homme;

être autonome et avoir un débit suffisamment important pour la collecte de quantités de poussières facilement pesables et analysables en routine;

être d'un côut suffisamment réduit;

mais surtout, pouvoir fonctionner dans toutes les positions et être insensible aux mouvements auxquels il peut être soumis. En effet, principalement dans le domaine minier ou l'homme est appelé souvent à travailler dans des conditions difficiles, courbé, à genoux, voire allongé, à proximité des sources de poussières, il est indispensable de pouvoir mesurer son exposition, étant entendu que les ouvriers qui sont amenés à travailler dans ces conditions sont généralement les plus exposés.

LE CIP 10

L'appareil est de petites dimensions (hauteur 170 mm, largeur max. 70 mm, épaisseur 25 mm) permettant d'être facilement glissé dans une petite poche de veste ou dans cele d'un baudrier.

Il comporte:

une tête proprement dite de prélèvement;

un boitier renfermant les éléments fonctionnels (moteur, batteries, circuit de régulation).

L'appareil effectue un prélèvement sélectif à deux étages.

Le premier étage ou sélecteur capte les plus grosses particules qui sont normalement arrêtées dans les voies respiratoires supérieures (nez, bouche, larynx, trachée). Il comporte pour cela un petit impacteur à mousse, suivi d'un bloc sélectif également en mousse de polyuréthane.

Le deuxième étage retient les particules respirables issues du premier, c'est-à-dire celles qui sont susceptibles de pénétrer dans les alvéoles pulmonaires et de s'y déposer. Il comporte pour cela une coupelle rotative tournant à 7000 t/min., renfermant un tore en mousse de polyuréthane.

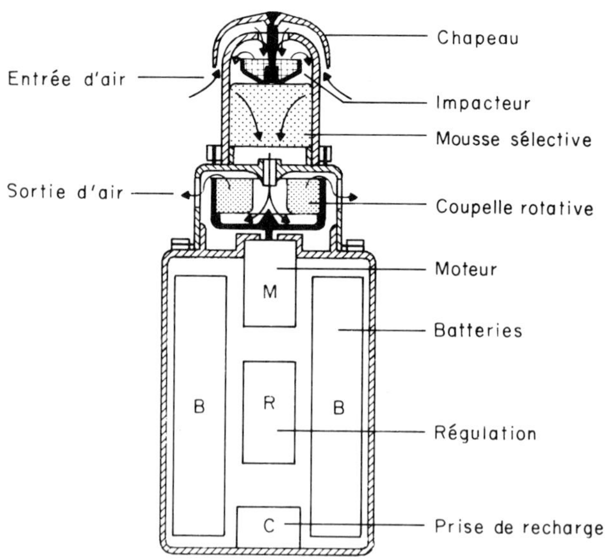

FIG.1. Le CIP 10.

Les plus fines particules qui sont normalement rejetées à l'expiration retournent également à l'atmosphère.

L'aspiration est réalisée par l'effet ventilateur de la coupelle rotative, sans nécessiter de systèmes déprimogènes complémentaires.

PRINCIPALES CARACTERISTIQUES

Consommation énergétique

La consommation énergétique de ce principe d'aspiration et de captation est particulièrement faible : 0.12 watt, ce qui a permis de concevoir ainsi un appareil léger : 300 grammes seulement et à haute performance : débit : 10 l/min. Autonomie : 20 heures.

Plan de la réalisation

Sur le plan de la réalisation de l'appareil une forme compacte et aveugle a été retenue pour la sécurité des prélèvements:

- il n'y a pas en particulier d'interrupteur apparent. La mise en route et l'arrêt de l'appareil s'effectuent à l'aide d'un aimant;
- les orifices d'entrée sont protégés par un chapeau évitant toute tentation de perturbation du prélèvement par les ouvriers (introduction de poussières à la main par exemple). Le chapeau a également pour effet de protéger le prélèvement contre les agents extérieurs indésirables (gouttes d'eau ou très grosses poussières):
- l'appareil comporte en outre des vis spéciales de blocage de la tête, destinées à empêcher son ouverture aux postes de travail;
- l'ensemble de l'appareil a enfin été réalisé en matière plastique moulée, antistatique, chargée de fibres de verre, particulièrement résistante. Il répond aux normes de sécurité les plus sévères concernant son utilisation dans les atmosphères explosibles.

Plan du fonctionnement

Sur le plan du fonctionnement, le principe utilisé avec des mousses à larges mailles sans effet sensible de colmatage et la régulation de la vitesse de rotation du moteur lui donne une très bonne constance du débit jusqu'à des prélèvements importants en poussières (50 mg pour les fines et 150 mg pour les plus grosses). It peut en outre fonctionner dans toutes les positions, est insensible aux mouvements (ou chocs) auquel il peut être soumis, et présente de ce fait une très bonne sécurité de prélèvement même dans les cas les plus difficiles.

Plan de son utilisation

Sur le plan de son utilisation, il est normalement mis en route en début de poste par un responsable, remis aux ouvriers à surveiller et récupéré en fin de poste.

Le dépouillement des mesures (pesées, dosage des constituants) est effectué ensuite au laboratorie et l'appareil rechargé pour de nouvelles mesures.

Les filtres sont facilement lavables, pour la récupération des poussières par exemple, et réutilisables après séchage pratiquement indéfiniment. Il n'y a pas lieu de réapprovisionner à chaque fois de nouveaux filtres.

RESULTATS OBTENUS A CE JOUR

Une soixantaine d'appareils sont en service depuis plusieurs mois dans les houillères françaises dans lesquelles jusqu'à présent seule une surveillance des ouvriers avec des appareils CPM 3 placés à poste fixe était effectuée.

Les premières conclusions sont les suivantes:

bonne acceptation de port par les ouvriers;
bon comportement de l'appareil dans les conditions les plus difficiles;
détection de niveaux d'exposition élevés, inattendus pour certains postes de travail ayant déclenché la mise en oeuvre de moyens spécifiques de prévention;
réflexions nouvelles sur le concept d'une surveillance appropriée des expositions en liaison avec les observations médicales.

CONCLUSION

Le nouvel appareil CIP 10 récemment mis au point au Centre d'Etudes et Recherches de Charbonnages de France et commercialisé par la Société MSA de France semble donc bien répondre aux exigences de la mesure des expositions effectives des ouvriers aux poussierès dans tous le cas et en particulier le plus difficiles et je vous remercie de m'avoir permis de vous le présenter.

ESTIMATING PAST EXPOSURES TO AIRBORNE NICKEL COMPOUNDS IN THE COPPER CLIFF SINTER PLANT

J. STUART WARNER
Inco Ltd., 1 First Canadian Place,
Toronto, Ontario M5X 1C4 CANADA

INTRODUCTION

In the past, there was an increased incidence of lung and sinus cancer among workers in certain nickel refining operations. The agent(s) responsible for the carcinogenic effects have not been identified with any precision nor have dose-response relationships been estimated. This lack of knowledge complicates hazard communication and effectively precludes the efficient management of risk. To rectify this situation, the considerable epidemiological effort expended on the nickel industry must be matched by efforts to estimate the relevant historic exposure conditions.

Making these estimates will be difficult but, perhaps, not impossible. It will require thoroughness and insatiable curiosity. Even if numeric data are found, they will generally have been gathered for purposes different from those of the "environmental historian" and must be interpreted. Each aspect of the estimate must therefore be approached from several directions if the final picture is to be internally consistent.

The easier part of estimating past conditions is to form reasonable hypotheses as to which chemical forms of nickel were or were not likely to have been present. The likelihood of exposure to non-nickeliferous carcinogens must also be considered. A thorough knowledge of the metallurgical processes used is essential and can usually be found in the open literature as well as in company records. Chemical thermodynamics and even simple stoichiometric calculations can be informative. Although it requires greater effort, laboratory experiments can often be arranged to subject the substances of interest to the process conditions of the past and see how they react.

Estimating the airborne concentrations of these substances and their temporal and spatial distributions is more difficult and the answers less certain. Making such estimates requires detailed knowledge of how the various processes were physically implemented in the refinery. This includes the layout of major equipment, materials handling methods, dust control equipment, etc. This information can usually be obtained from company records but the men who worked in the plant are even more valuable sources of information.

All of the above techniques are being used to estimate the environmental conditions that existed in the Copper Cliff Sinter Plant (CCSP), a nickel refining operation in which there was an increased risk of respiratory cancer. The CCSP operated from Feb. 1948 until Feb. 1963. In subsequent epidemiological studies, Roberts et al. (1984) found the incidence of both lung and nasal cancer was significantly increased, even among men who had worked there for less than one year. The authors also noted that all of the sinus cancer deaths and 60 of the 65 lung cancer deaths found by the end of 1981 occurred in men who were first exposed before 1952. Although this striking difference in response may vanish as the CCSP cohort ages, it suggests that special attention be paid to the temporal variation of environmental conditions in this plant.

METHODS

Information about the CCSP was obtained by an extensive search of company records including: weekly, monthly and annual operating reports; metallurgical accounting reports; research reports; engineering records and drawings; maintenance records; and purchasing records. Records maintained by the Ontario government were also examined. Most valuable of all were the discussions with men who had worked in the CCSP. Their interpretations and perspective made the recorded information even more useful. Visits to the remains of the CCSP were also helpful.

Laboratory experiments were conducted to see whether certain nickel compounds might have been formed unexpectedly in the CCSP. The performance of equipment once used to sample dusty air in the CCSP was compared with that of personal gravimetric samplers.

FINDINGS AND DISCUSSION

Chemical forms of nickel in dust

In the CCSP, various sulfur-containing forms of nickel were oxidized to nickel oxide (NiO) on traveling grate sintering machines. The principal feeds were fine flotation concentrates of nickel subsulfide (Ni_3S_2) containing differing amounts of copper sulfide. Small amounts of reverts from other plants were also processed including nickel sulfate ($NiSO_4$) and basic nickel carbonate ($Ni(OH)_2 \cdot NiCO_3$). The charge to the sintering machines was normally **not** sintered completely so a significant fraction of the feed was discharged hot and dry. This problem was particularly severe in the first four years.

The material discharged from the machines was crushed twice, conveyed and screened while dry and at temperatures over 500 °C. Even fully-sintered material generated NiO dust in these operations. The presence of unsintered material greatly aggravated the situation.

Laboratory experiments indicated that $Ni(OH)_2$ was not formed by cooling sinter with water. The possibility that $NiSO_4$ was formed from Ni_3S_2 under certain conditions is still being investigated.

Dust from a single sample of air taken on the operating floor in 1960 was chemically analyzed. Because the CCSP was so dusty, the amount of nickel and copper lost in the dusty air escaping from the roof monitors was measured from time to time. A few analyses of deposited dust (on snow, the CCSP roof, etc.) were also found in the records and 13 "posthumous" samples were taken from elevated sections of the CCSP for analysis. Reconciliation of the data is not yet completed but it appears that NiO, Ni_3S_2 and $NiSO_4$ were each present in the dust, $NiSO_4$ being the least concentrated.

Concentrations of airborne dust

A single 40-hour sample of dusty air on the operating floor gave the concentration of total dust as 46.4 mg/m^3 in late 1960. The reports of metal losses from the roof monitors also included the flowrate of air so the concentrations of dust at the monitors could be calculated. Since nearly all the air exited the plant through the monitors, the monitor samples may be used as estimates of the working environment. In fact, they are probably underestimates as the dusty air leaving the monitors was often diluted with clean air from the windows above the operating floor. It is also known that large amounts of coarser dusts recirculated at lower elevations in the CCSP.

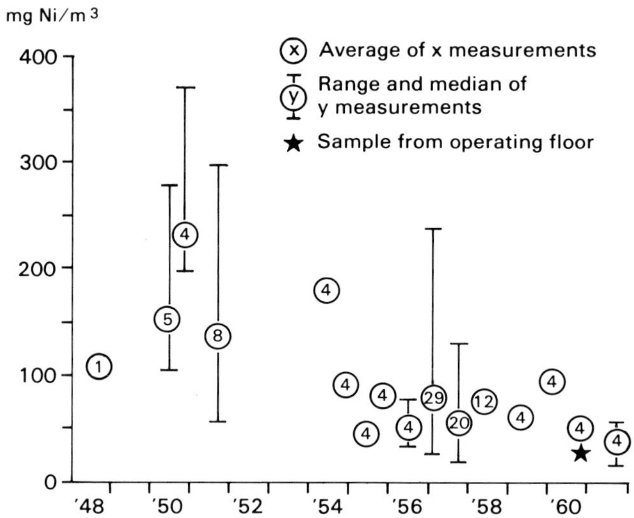

FIG 1 Concentrations of nickel in dusty air leaving the roof monitors of the Copper Cliff Sinter Plant

Fig 1. shows the concentration of airborne nickel was lowered significantly with time but was always extremely high by today's standards. The long term trend correlates reasonably well with improved operation of the sintering machines and with the installation of dust control equipment. The equipment used to sample the dust at the CCSP monitors is now being run side by side with modern personal gravimetric samplers to see whether the results are directly comparable.

REFERENCE

Roberts R.S., Julian J.A. and Muir D.C.F. (1984) These proceedings, p.207

461 645

MORTALITY ANALYSIS IN A CANADIAN SINTER PLANT: A COMPARISON OF TWO COHORTS BASED ON YEAR OF FIRST HIRING

David C.F. MUIR, James A. JULIAN, and Robin S. ROBERTS

Health Sciences Centre, McMaster University,
Hamilton, Ontario, L8N 3Z5 CANADA

INTRODUCTION

An historical prospective mortality study has been undertaken of male workers employed by INCO in its Ontario operation. Initial results of this major epidemiological survey have been reported (Roberts et al 1980) and full details will be published shortly. In summary the findings confirmed that there was a high mortality due to nasal and lung cancer in the sinter plant operations.

The sinter plant located at Copper Cliff, Sudbury began operating in 1948 and was replaced by alternative production methods in 1963. Special efforts to reduce environmental exposure of workers were made and these probably resulted in lower airborn dust levels in the later years of the operation. The effects of these hygiene improvements on cancer mortality are examined in the present paper by dividing the workforce into two groups according to the year of first sinter plant exposure and by comparing the mortality rates in the two groups. Environmental dust exposure data were not available for analysis but are reported separately at this conference (Warner, 1984). However, information from ventilation engineers at INCO suggested that dust levels began to fall after 1952 at which point the initial process control difficulties had been solved. Special attention in the present report is given to two practical issues. Firstly, can it be shown that the ventilation improvements designed to reduce dust exposure to workers were effective in reducing cancer mortality? Secondly, were they sufficiently effective to reduce cancer mortality to the point where it could not be distinguished from that in the normal population.

METHODS

All workers with any exposure at the Copper Cliff sinter plant (at Sudbury) from its opening in 1948 to its closure in 1963 were included in the cohort for study. The mortality experiences of the cohort were examined from 1950, when death certificates were first made available for computer linkage in Canada through to 1981. The cohort was divided into two. The first contained 1,180 workers who first experienced sinter plant exposure before 1952. The second contained 672 workers whose exposure was restricted to the years 1952 or later. Details of the two groups are shown in Table I and II which also illustrate the numbers of deaths recorded. The number of deaths due to nasal cancer was too small for analysis which was accordingly restricted to a comparison of lung cancer death rates in the two subcohorts. Such an analysis must take account where necessary, of the differences in the means of age at entry, years of exposure and of followup of the two groups. This was achieved by stratifying for duration of exposure and by restricting followup to 29 years from first exposure so that it was of equal duration in the two groups. Only lung cancer deaths occuring 15 or more years since first sinter plant exposure were included in the analysis in order to ensure that there was an adequate latent period since first sinter plant exposure. The numbers were too small to permit an examination of the small difference in the mean age of first exposure between the two groups, but this is not felt to be an important determinant of lung cancer rates.

TABLE I. Copper Cliff Sinter Plant 1948-1963

	First Sinter Plant Exposure	
Count	Before 1952	1952 or later
Cohort size	1180	672
1950-1981 deaths:		
- all causes	255	51
- lung cancer	60	5
- nasal cancer	5	0

TABLE II. Copper Cliff Sinter Plant 1948-1963

	First Sinter Plant Exposure	
Mean	Before 1952	1952 or later
Age at entry	27	24
Years of exposure	2.4	1.8
Years of followup	28	25

TABLE III. Copper Cliff Sinter Plant 1948-1964 (period in view 1950-1981)

Lung Cancer Mortality (15-29 years since first exposure)

Years of Sinter Exposure	Sinter Plant Exposure					
	Before 1952			1952 or later		
	OBS	EXP	SMR	OBS	EXP	SMR
<5	23	7.76	296	3	2.15	139
5-9	11	0.64	1730	1	0.30	339
10 +	18	1.51	1197	1	0.23	431
All	52	9.90	525	5	2.68	187
95% CI on SMR			(460,590)			(60,437)

RESULTS

The results of the analysis are shown in Table III. Workers first exposed prior to 1952 have experienced a mortality rate from lung cancer five times that of the global population. The true relative risk probably lies between 4.6 and 5·9 and is strongly related to the duration of exposure. By contrast the smaller group of men first exposed in 1952 or later have experienced an approximately twofold increase in the risk of dying from lung cancer. The observed SMR of 187 for this group is statistically different ($\rho<0.05$) from the SMR of 525 recorded by the pre 1952 group indicating a reduction in risk over time. The confidence interval for the post 1952 group is wide, however, indicating that the first risk may still be as high as an SMR of 437 or as low as 60. There is then the possibility that the lung cancer risk has returned to that of the global public.

DISCUSSION

Clearly, the findings of this analysis provide a reliable answer to the first question raised in the introduction. The risk of mortality from lung cancer has fallen sharply between men first exposed prior to 1952 and those exposed subsequently. The observed reduction in risk of 65% can be reasonably attributed to improved hygiene conditions in the plant. The relative paucity of data in the post 1952 group has made it difficult to reliably answer the second question of whether the risk has been eliminated. The observed lung cancer SMR of 187 is consistent with a wide range of time underlying risk including no excess risk. However, the fact that earlier workers were clearly at risk and observed lung cancer risk in subsequent workers was elevated (even if not conventionally significantly so) takes us, at this stage, to the prudent conclusion that a modest lung cancer risk remained in the Copper Cliff Sinter Plant after 1952. Their exposure in the years following 1952 were probably of gradually falling intensity (Warner, 1984) and it may be that workers having first exposure in even later years of the sinter operation did not have demonstrably increased death rates. Numbers of deaths and years of followup in our analysis are inadequate to answer this question at present, but may become so in due course.

REFERENCES

ROBERTS, R.S., JULIAN, J.A., SHANNON, H.S. AND MUIR, D.C.F.(1980) Mortality studies in Ontario of nickel workers: the INCO/ JOHC study in Nickel Toxicology (Eds S.S. Brown and F.W. Sunderman, Jr.), pp. 27-30. Academic Press, London.

WARNER, S. (1984) These Proceedings, pp. 203

461 646

INCIDENCE DES CANCERS RESPIRATOIRES CHEZ LES TRAVAILLEURS D'UNE ENTREPRISE D'EXTRACTION ET DE RAFFINAGE DU NICKEL ET DANS LA POPULATION GENERALE EN NOUVELLE-CALEDONIE

Pâquerette GOLDBERG, Michel BLANC, Rebecca FUHRER, Nereo SEGNAN, Jean-Marc BRODEUR, Annette LRCLERC, Denis FRANCOIS, William DAB, Catherine GODARD et Marcel GOLDBERG

INSERM U 88-GERSS, 91 bd de l'Hôpital, F-75634 Paris Cèdex 13 France

INTRODUCTION

L'objectif de cette étude est d'apporter une réponse globale sur une relation possible entre le processus d'extraction et de raffinage du nickel et les cancers des voies respiratoires parmi les travailleurs de la Société Le Nickel-SLN en Nouvelle-Calédonie. Pour cela, on compare l'occurrence des cancers entre les travailleurs et anciens travailleurs de la SLN et la population générale n'ayant jamais travaillé dans cette entreprise. L'indice utilisé est le taux d'incidence annuel, estimé pour chacune des deux populations comparées.

MATERIEL ET METHODE

Le modèle épidémiologique le mieux adapté pour une analyse de ce type est une étude de cohorte historique. Le manque total ou partiel des données médico-démographiques nécessaires n'a pas permis de suivre ce modèle qui suppose en outre que la population des travailleurs soit très petite devant la population générale, ce qui n'est pas le cas pour la Nouvelle-Calédonie.

On a estimé, pour une période d'observation de quatre années 1978-1981, des taux d'incidence annuels par classe d'âge et pour chacune des deux populations : hommes travaillant ou ayant travaillé à la SLN (TE) et hommes n'ayant jamais travaillé dans cette entreprise (PG) et pour les deux pathologies mises en cause dans l'industrie du nickel : cancers pulmonaires (CP) et cancers des voies aériennes supérieures (CVAS).

Les numérateurs des taux d'incidence ont été établis grâce à une recherche exhaustive des cas de cancer respiratoire en Nouvelle-Calédonie incidents pendant la période d'observation. Cette recherche s'est déroulée selon plusieurs étapes successives.

1. Définition de critères a priori permettant de suspecter un cas
2. Etablissement de la liste des sources de cas potentiels (16 sources)
3. Recherche de tous les dossiers médicaux dans ces sources correspondant aux critères
4. Collecte des informations médicales et établissement d'un dossier standardisé
5. Certification par un comité d'experts médicaux réuni à Paris, selon une procédure standardisée et à l'aveugle, c'est-à-dire sans identification du sujet et sans la connaissance du diagnostic préalable ou de son absence.

Le protocole de certification a permis de classer chaque cas potentiel en positif, négatif, ou douteux. Tous les dossiers douteux ont fait l'objet d'une double certification et ont pu être classés en douteux positifs qui ont été retenus pour l'étude et douteux négatifs qui ont été exclus.

Pour les cas masculins inclus l'appartenance à la population TE ou PG a été déterminée de deux façons : recherche dans les archives de l'entreprise et entretien toutes les fois que cela a été possible avec le sujet ou un membre de sa famille. La concordance entre ces deux procédés a été parfaite dans tous les cas où l'entretien a eu lieu avec le sujet. Les résultats figurent au Tableau I.

Les dénominateurs des taux d'incidence sont les effectifs de population correspondant au Tableau I. La population TE des travailleurs et anciens travailleurs de l'entreprise, vivant en Nouvelle-Calédonie pendant la période d'observation (1978 à 1981), a été reconstituée à partir d'une étude démographique basée sur le fichier des embauches de l'entreprise depuis 1930 et sur un échantillon de personnes embauchées par l'entreprise de même composition d'âge que les cas.

Deux hypothèses opposées ont été retenues pour estimer les effectifs de la population TE moyenne pendant la période d'observation : la première hypothèse (hypothèse basse) correspond à une faible mortalité et une faible émigration parmi les travailleurs embauchés par l'entreprise depuis 1930, la deuxième hypothèse (hypothèse haute) correspond à une forte mortalité et une forte émigration parmi ces mêmes travailleurs.

La population totale des hommes vivant en Nouvelle-Calédonie pendant la période d'observation a été déterminée à partir du recensement général de la population de 1976 et de données de projection.

TABLEAU I. Effectifs des cancers des voies respiratoires pour la période d'observation et en fonction de l'appartenance à la population.

TE : travailleurs et anciens travailleurs de l'entreprise
PG : population générale
CP : cancer pulmonaire
CAVS : cancer des voies aériennes supérieures

	CP		CVAS	
Classes d'âge	TE	PG	TE	PG
25-29	-	-	-	1
30-34	1	1	-	-
35-39	1	1	-	-
40-44	2	7	-	1
45-49	7	9	6	2
50-54	5	7	6	6
55-59	6	13	2	5
60-64	6	11	-	8
65-69	4	11	-	4
Total	32	60	14	30

La population générale PG des hommes vivant en Nouvelle-Calédonie pendant la période d'observation et n'ayant jamais travaillé dans l'entreprise étudiée a été calculée par différence entre la population totale et la population TE :

Population totale = TE + PG

Les résultats des dénominateurs sont donnés au Tableau II.

RESULTATS

Les taux d'incidence annuels cumulatifs sur les classes d'âge ont été calculés ainsi que les écarts types et les intervalles de confiance à 5 % sous l'hypothèse de normalité. Les résultats sont donnés au Tableau III.

TABLEAU II. Estimations des populations moyennes

		Hypothèse basse		Hypothèse haute	
Cl. d'âge	Pop. totale	Pop. TE	Pop. PG	Pop. TE	Pop. PG
25-29	5374	742	4632	594	4780
30-34	5046	1429	3617	1145	3901
35-39	4492	2192	2300	1756	2736
40-44	3967	1667	2300	1334	2633
45-49	3332	1590	1742	1371	1961
50-54	2818	1430	1388	1231	1587
50-59	2162	837	1325	718	1444
60-64	1416	535	941	483	993
65-69	1298	353	945	315	983
Total	29965	10775	19190	8947	21018

TABLEAU III. Taux d'incidence annuels cumulatifs

	CP				CVAS			
	Hyp. basse		Hyp. haute		Hyp. basse		Hyp. haute	
	TE	PG	TE	PG	TE	PG	TE	PG
Taux cumulatif	5,00	5,89	5,70	5,44	1,29	2,99	1,50	2,76
Ecarts types	1,03	0,79	1,17	0,74	0,36	0,57	0,42	0,53
Interval de conf.	2,94	4,21	3,36	3,96	0,57	1,85	0,66	1,70
à 5 %	-7,06	-7,45	-8,04	-6,92	-2,01	-4,13	-2,34	-3,82

En supposant que le taux d'incidence annuel cumulatif suive une loi normale, on a comparé les deux populations TE et PG pour ce taux. Les risques relatifs et les résultats du test sont donnés au Tableau IV.

La seule différence observée concerne les cancers des voies aériennes supérieures dans l'hypothèse démographique d'une faible émigration et mortalité, et est dans le sens où la population de travailleurs TE exposée au nickel a un risque moindre pour cette pathologie que le reste de la population.

TABLEAU IV. Risques relatifs de la population de travailleurs par rapport à la population générale pour
CP : cancers pulmonaires
CVAS : cancers des voies aériennes supérieures

	CP		CVAS	
	Hypot. basse	Hypot. haute	Hypot. basse	Hypot. haute
Risques	0,85	1,05	0,43	0,54
relatifs	NS	NS	$p<0,05$	NS

DISCUSSION

Cette étude comporte deux biais potentiels principaux. Premièrement, les dénominateurs des taux d'incidence ont été estimés à partir d'hypothèses démographiques sur l'émigration et la mortalité. Des approximations ont été faites et ont pu influer sur les résultats qui sont sensibles, en particulier à la composition par classes d'âge des populations étudiées. Deuxièmement, des personnes exposées au nickel dans d'autres entreprises que la SLN ont été comptées (numérateurs et dénominateurs) avec la population générale car aucune donnée actuelle ne permet d'estimer la population correspondante. L'effet certain que cela entraîne n'est pas mesurable sans recueil de données complémentaires.

461 647

CANCER DES VOIES RESPIRATOIRES PARMI LES TRAVAILLEURS D'UNE ENTREPRISE D'EXTRACTION ET DE RAFFINAGE DU NICKEL EN NOUVELLE-CALEDONIE : UNE ETUDE CAS-TEMOINS AU SEIN D'UNE COHORTE.

Marcel GOLDBERG, Rebecca FUHRER, Jean-Marc BRODEUR, Pâquerette GOLDBERG, Nereo SEGNAN, Annette LECLERC, Jean-François CHASTANG, Renée BOURBONNAIS et Denis FRANCOIS

INSERM U 88-GERSS, 91, bd de l'Hôpital, F-75634 PARIS Cédex 13 France

OBJECTIFS

L'étude qui est rapportée ici avait comme objectif de tenter de déterminer s'il existe, chez les travailleurs de la Société Le Nickel (SLN) en Nouvelle-Calédonie, un lien entre cancer de l'appareil respiratoire et niveau d'exposition aux produits de l'extraction et du raffinage du nickel.

Cette étude a été menée en parallèle avec l'étude d'incidence visant à comparer la population SLN et la population générale de Nouvelle-Calédonie qui a été rapportée à cette même conférence (cf. P. Goldberg et al., 1984), et certains points du protocole sont communs aux deux études.

MATERIEL ET METHODES

Le protocole est celui d'une enquête cas-témoins au sein d'une cohorte.

La cohorte considérée est celle de l'ensemble des hommes ayant travaillé à la SLN depuis 1930 jusqu'à 1977, quelle que soit la durée de l'emploi.

Les cas sont la totalité des cas masculins de cancers respiratoires incidents dans la cohorte du 1.1.1978 au 31.12.1981 : cancers du poumons et cancers des voies aéro-digestives supérieures (VADS). Il s'agit du sous-ensemble des cas survenus en Nouvelle-Calédonie pendant cette période qui ont travaillé à la SLN : la procédure de recherche et de certification de ces cas a été décrite avec l'étude d'incidence

Les témoins ont été choisis au sein de la cohorte. Les témoins ont été définis comme des personnes appartenant à la cohorte et étant à risque d'avoir un cancer respiratoire pendant la période d'observation (1978-81), c'est-à-dire vivants et présents en Nouvelle-Calédonie au 1.1.1978. Pour chaque cas, les 2 personnes de la cohorte les plus proches par l'âge et les 2 personnes les plus proches par la date d'embauche ont été sélectionnées.

Des informations individuelles concernant chaque sujet ont été recueillies à différentes sources :

. enquête par interview auprès des sujets (ou d'un autre répondant en cas d'impossibilité : décès, départ, etc.). Un questionnaire a été passé pour recueillir notamment les informations suivantes : caractéristiques démographiques et socio-familiales, histoire d'habitation, histoire professionnelle, consommations détaillées de tabac et d'alcool.

. archives de la SLN pour reconstituer de façon précise la carrière professionnelle au sein de la SLN : suite des postes occupés avec dates de début et de fin de chacun d'eux.

. "Comité d'évaluation de l'exposition", composé de personnes désignées par la Direction et les organisations syndicales. Ce Comité avait plusieurs objectifs : i) définir la liste des nuisances professionnelles existant au ayant existé à la SLN depuis 1930 : 14 nuisances professionnelles ont été définies (Tableau I). ii) reconstituer l'histoire des modifications technologiques apportées aux processus de production depuis cette date, et définir des ensembles de postes exposés de façon homogène aux mêmes nuisances pendant une certaine période : ces ensembles ont été appelés des "postes-périodes", et le Comité en a défini 105 pour la raffinerie et 24 pour les mines. iii) attribuer à chaque "couple poste période/nuisance" une note quantitative de 0 à 10, reflétant le niveau relatif de pollution du poste par la nuisance considérée ; pour l'attribution de ces notes, une méthode de type DELPHI a été utilisée : cette méthode, qui vise à recueillir l'opinion d'un groupe "d'experts", repose sur l'anonymat des réponses individuelles des experts, un processus d'itération et de rétroaction contrôlée, et une réponse statistique de groupe. De cette façon, une valeur cumulée d'exposition pour chaque nuisance a pu être établie individuellement pour tous les sujets de l'étude, en sommant les valeurs de nuisance attribuées aux différents postes occupés au cours de leur carrière à la SLN en fonction des durées et des périodes correspondantes.

TABLEAU I. LIST OF OCCUPATIONAL HAZARDS

CARBON	"INDUSTRIAL" ASBESTOS
TOTAL DUST	COMPOUND GASES
NICKEL DUST	HEAT
RAW ORE	WOOD DUST
ROASTED ORE	ROAD AND WORK SITE DUST
TAR	PERFORATION DUST
SILICA	DIESEL ENGINES GASES

RESULTATS

Le groupe étudié se compose de 186 personnes, cas et témoins ; 158 interviews ont été complétées (Tableau II)

TABLEAU II. STUDY GROUP

	CASES		CONTROLS	TOTAL
N	49	LUNG : 34 URT : 15	137	186
COMPLTD. INTERVIEWS	47 (96 %)		111 (81 %)	158 (85 %)

MEAN LATENCY TIME : LUNG : 25.5 YRS
URT : 24.3 YRS

L'exposition aux tiers-facteurs individuels (âge, tabac, alcool) a été comparée pour les 3 groupes de sujets (témoins, cas poumon, cas VADS). Le Tableau III montre que les consommations de tabac et d'alcool sont plus élevés chez cas.

TABLEAU III. CONFOUNDING VARIABLES

	CONTROLS	LUNG	URT	TEST C/L	TEST C/U
MEAN AGE	54.7	56.3	52.9	NS	NS
SMOKING					
% smok	75.5	100	100	- $p < 0.001$	-
Nmb. cig	205,000	429,000	422,000	$p < 0.01$	$p < 0.05$
Gr. ALC	835,000	2,960,000	1,000,000	$p < 0.05$	NS

La répartition des principales variables professionnelles entre les 3 groupes montre que, paradoxalement, les témoins semblent plus exposés que les cas. Le Tableau IV montre que la durée moyenne d'exposition moyenne, la fréquence et la durée du travail dans des postes directement liés à la production sont plus élevés parmi les témoins alors que les cas sont plus nombreux dans les emplois de bureau (tendances non significatives).

TABLEAU IV. MAIN PROFESSIONAL CHARACTERISTICS

	CONTROLS (n = 137)	LUNG (n = 34)	URT (n = 15)	TEST C/L	TEST C/U
LENGTH EMPL. (IN YEARS)	13.9	11.3	11.6	NS	NS
% WORKERS					
Production	54	52	44	NS	NS
Offices	9	18.5	22	NS	NS
% TIME					
Production	58	37.5	39	NS	NS
Offices	8.7	27.2	25.9	NS	NS

L'étude des variables individuelles d'exposition cumulée aux 14 nuisances spécifiques par le Comité d'Evaluation de l'Exposition confirme cette observation: quelque soit la façon de comparer les témoins et les cas, on ne trouve pratiquement aucune différence significative entre groupes et sujets, et sauf exception, ce sont les témoins qui sont les plus exposés.

Le Tableau V résume ces constatations: pour les 4 nuisances les plus importantes, les valeurs moyennes d'exposition individuelle cumulées sont presque toutes plus élevées pour les témoins.

TABLEAU V. MAIN HAZARD EXPOSURE (MEAN CUMULATED VALUES)

HAZARD	CONTROLS (n = 137)	LUNG (n =34)	URT (n = 15)	TEST GLOBAL	TEST C/L	TEST C/U
TOTAL DUST	33.5	21.5	27.2	NS	NS	NS
NICKEL DUST	17.3	9.9	12.2	NS	NS	NS
RAW ORE	46.3	23.5	33.6	$p<0.05$	$p<0.01$	NS
ROASTED ORE	24.7	14.2	25	NS	NS	NS

CONCLUSION

Alors qu'on retrouve facilement le rôle des facteurs de risque traditionnels (tabac,alcool), il n'a pas été possible de mettre en évidence un rôle des facteurs professionnels spécifiques.

Cependant cette étude manque de puissance statistique, et il est prévu de la continuer sur une plus longue période pour augmenter l'effectif des sujets.

REFERENCE

Goldberg, P., Blanc, M., Fuhrer, R., Segnan, N., et al. These Proceedings, pp. 211

461648

TOXICOLOGICAL AND OCCUPATIONAL MEDICAL INVESTIGATIONS CONCERNING NICKEL-EXPOSURE IN DIFFERENT INDUSTRIAL AREAS IN THE FRG

HANS-JÜRGEN RAITHEL, WOLFRAM KRESS, KARL-HEINZ SCHALLER, DIETER WELTLE and HELMUT VALENTIN

Institute of Occupational and Social Medicine,
University of Erlangen-Nürnberg,
Schillerstr. 25, D-8520 Erlangen, FRG

INTRODUCTION

Nickel and its compounds can cause acute disease symptoms in humans and may have serious long term effects. Besides the allergic contact eczema known for many years the carcinogenic effect of nickel (and especially its derivatives of low solubility) are the subject of much recent scientific discussion. In the period from 1968 until the end of 1983 13 cases of bronchial and 3 cases of nasal carcinoma were recognized as nickel-induced occupational diseases in West Germany. The cases refer to workers who were employed by two companies refining predominantly nickel residues.

A critical review of the scientific literature, however, showed that the situations of exposure in the different industrial areas has to be viewed in very different ways. Moreover, there is a substantial lack of information about the actual exposure to nickel in many working areas. Therefore the purpose of this report is to review occupational medical and toxicological investigations in various industries.

MATERIALS AND METHODS

A total of 649 persons from four different industrial areas was examined. There were 54 women and 42 men of a company producing nickel-cadmium batteries; 44 men of a company refining nickel electrolytically; 60 men of a factory producing nickel-alloys and pure nickel in powder form; 449 workers from the glass-industry who where occupied with building or repairing nickel-armoured hollow glass moulds.

94 men without occupational nickel exposure served for comparison purposes.

The examinations were carried out according to the strain-stress concept. In terms of strain, it included the quantitative determination of the external and internal (biological monitoring) nickel exposure at different work sites.The nickel analyses of the samples were carried out by electrothermal AAS.

In terms of stress, examinations of lungs, kidneys and skin were central, with special regard to the pathophysiology as well as the kinetics of absorption and excretion.

RESULTS AND DISCUSSION

Measured values concerning the external nickel exposure were available from the industrial area of the nickel refinery, from the alloy production, and especially form the glass industry. In the last mentioned industry we had in addition the possibility to compare values from 1981 with those from the spring of this year. Between these two dates, work hygiene measures had been improved to some extent, especially in firms A and C of the glass industry with respect to improvements in the exhausters.

In the nickel industry we found a median value of 50 μg/m³ within the melting area and a median value of 300 μg/m³ within the section of powder metallurgy. In this area the technical guiding concentration (TGV-value) of 500 μg/m³ for nickel and its compounds valid for Western Germany (calculated as nickel concentration in the total dust) was exceeded at 2 out of 8 measuring stations.

Within the glass industry the values varied - depending on the efficacy of the exhauster - between the limit and a maximum of 604 μg/m³ for flame spraying. When comparing the values with those of 1981, we could certainly establish a decrease in the external exposure at firm A. This may be explained by the improvement and the consequent effect of work hygiene measures.

The creatinine-related nickel-concentrations in after shift urine samples of a total of 138 workers, in four glass factories showed that the exposure limit of 25 μg/g creat. was exceeded in all firms. The situation of internal exposure, however, is very different. The highest average values were found in firm C.

For 17 smaller glass works, we were able to establish quite different internal exposure situations. When comparing the values found in firm A to D with those measured in 1981 (Table 1), an obvious decrease of the median values (15.5 to 5.0 μg nickel/g creat.) was seen for firm A, whereas with firm C an increase of the median creatinine-related values from 24 to 28.5 μg nickel/g creat. is noteworthy.

TABLE 1 Medians,(66.6-percentiles) and % above 25 µg Ni/g Cr. for renal Ni-Excretion and Ni in plasma.

glass industry factory	1981 Ni-U (µg/g Creat.)	1984 Ni-U (µg/g Creat.)	% above (µg/g Cr.)	Ni-Plasma (µg/l)
A	15.5 (6.5 - 43.0)	5.0 (1.0-15.0)	16	1.23 (0.63-1.86)
B	7.0 (2.5 - 10)	5.0 (2.5-21)	16	1.58 (0.88-2.42)
C	24.0 (3.8 - 112)	28.7 (16 -88)	61	2.45 (1.58-4.24)
D	6.0 (2.5 - 17.5)	4.0 (0.5-22.5)	21	0.77 (0.25-1.58)
Non-exposed workers (N=94)		1.1 (0.3- 2.0)		

The workplace-related evaluation of the nickel-values in urine shows the highest burden (median 37 µg/g creat.)for the group of welders (flame sprayers). By contrast, we found relatively low internal exposure for the mechanically working polishers and chasers (3.3 µg/g Creat.) and a medium exposure for the combined working welders and mechanics (17 µg/g Cr.).
In the course of the time the nickel urine concentrations of the welders have markedly increased from 1981 to 1984. For the mechanicians and for the combined working persons there was no major change.
The different exposure situations were confirmed by the findings for nickel plasma concentrations.
A significant correlation between Ni-concentrations in pre-shift urine and plasma is to be seen. The correlation between the after-shift urine values and the nickel concentrations in the plasma samples, however, is not as good. In this case it is remarkable, that short term changes of the external exposure situation can be described most reliably by a corresponding increase or decrease in renal nickel elimination.
Finally we compared the nickel concentrations of pre-shift urine samples with those of after-shift urine samples.In principle, the values obtained from after-shift urine were higher. Beyond that a significant correlation resulted from the creatinine-related values as well as from the volume-related ones.
The median values of the nickel-concentrations obtained in the other industrial areas, their 5- and 95-percentiles and their minimal and maximal values are shown in Table 2.
It can be seen that in addition to the glass industry, comparably high internal exposures can be found in the electrolytic nickel-refineries.

Table 2: Medians, (5/95 percentiles) and min./max ranges for the renal Ni-excretion of workers in different

Occupational group	N	Ni-U (µg/g Cr.)	4./95.-percentile	min./max. range	% above 25 µg/g Cr.
Ni-Cd battery workers	96	3.3	1.1 - 69	0.4 - 121	8
Ni-Refinery workers (electrolyt.process)	44	14.8	3.8 - 58	2.5 - 63	29
Ni-Refinery production of alloys and Ni-powder	59	2.6	0.8 - 36	0.5 - 52	5

The occupational medical examinations had the following results:
With regard to the smoking habits of the examined persons, there were no pathological results of lung function analyses kel dusts or aerosols. Moreover, the radiological thorax survey by X-ray technique did not show any pathological changes which could be interpreted as nickel induced. Above all no indication of a malignant process could be characterised.
The parameters collected as indicators of a possible nephrogenic injury - creatinine in the serum, total protein in the urine and β_2-microglobulin fraction (as an indicator of a tubular injury) - were within the normal range, too. Finally we could not detect any cutaneous Ni-sensitization.

CONCLUSIONS

1. For the occupational groups examined we have to recognise distinct external and internal exposures. The dust measurements as well as the determination of the nickel concentrations in urine and plasma resulted in values above the current TGV-values.
2. Increased in urinary nickel elimination have to be into be seen as a relevant measure of exposure.
4. At present no indication of nickel-induced debility or disease could be found.
5. Working in the nickel-using glass industry or in electrolytic nickel refineries should be monitored by occupational medical examinations.

461649

HEALTH SURVEILLANCE OF SLN WORKERS

Jean MEININGER and B. DAUTZENBERG
Société Métallurgique Le Nickel-SLN
75755 Paris Cédex 15 France

INTRODUCTION

Various rules and regulations govern an employer's obligations with respect to occupational medicine in France.

Since 1946, every plant is required to provide health surveillance of its employees.

For this purpose, the doctor must spend a minimum time of:

- one hour monthly per twenty office workers,
- one hour monthly per fifteen production workers,
- one hour monthly per ten workers exposed to substances designated as toxic, such as arsenic, lead, etc.

Doctors of occupational medicine must have a relevant degree which involves two years of specialization.

Their role is specifically limited to health surveillance of personnel and prevention of occupational diseases and accidents in the workplace.

For industrial facilities with between two hundred and eight hundred employees, there must be a full-time nurse; for over eight hundred employees, two full-time nurses.

Occupational doctors must spend a third of their time in the workplace and be familiar with the working environment and conditions of exposure to designated substances.

SLN AND HEALTH SURVEILLANCE

Nickel is not a designated substance in France. Therefore, in a plant like SLN Sandouville employing some 200 production workers and 70 office workers, the doctor is required by law to spend at least 18 hours a month (14 + 4 respectively). The Sandouville plant doctor actually spends twenty-four hours a month.

In the New Caldedonia plant of SLN, which employs more than two thousand persons, there are two full-time doctors.

Their situation is slightly different, because they not only practice occupational medicine, but also serve on occasion as general practitioners for plant personnel.

In other respects, the legislation currently in force in New Caledonia encompasses practically the same obligations as in Metropolitan France.

The preventive health program for nickel workers in both Sandouville and New Caledonia involves:

1. a pre-employment examination, (but there has been very little hiring of new employees for several years);
2. examination of all employees at least once a year. This examination consists of:
   - medical history, including smoking habits, previous dermatitis, respiratory allergy, or nasal injuries;
   - physical examination.

   In addition to those examinations prescribed by law, we do:
3. an annual X-ray-film . If the doctor has the slightest doubt, he performs a standard X-ray. Blood tests are also performed with cell counts, creatinine, glucose, uric acid and cholesterol.
4. an oto-rhino-laryngological examination by a specialist who has visited the plant once a week since 1980. This physician classifies his clinical results as follows:

   0 = normal
   1 = diffuse inflammation and flush of the mucosis
   2 = flush and leucoplastic aspect
   3 = aspect justifying a biopsy.

   Up to now, there have been no cases of class 3, and in the past year, only four of class 2, all heavy smokers.
5. biological monitoring by urine and/or plasma nickel analysis. This was initiated in Sandouville in 1979, and in New Caledonia in 1981.

   In Sandouville, the following numbers of tests were performed:

- one hundred and six urine analyses in 1979
- two hundred and eighteen urine analyses in 1980
- one hundred and ninety-two urine analyses in 1981.

However, there have been many doubtful results because of difficulties in collecting uncontaminated samples. Furthermore, as there was no rejection of urine specimens with specific gravity values outside the 1016-1026 range, results did not correlate with nickel air levels. Therefore, the analysis program was temporarily stopped, but it is hoped to start it up again in 1985 with appropriate modification.

The Doniambo plant in New Caledonia has benefited from the experience gained in our Sandouville program. Moreover, the conditions there were better (e.g. a larger medical department located near but not inside the works) and the results of urinary nickel analyses were more credible (Fig. 1). Thus the two year's results are very comparable and consistent with current air levels of predominantly insoluble nickel in this plant.

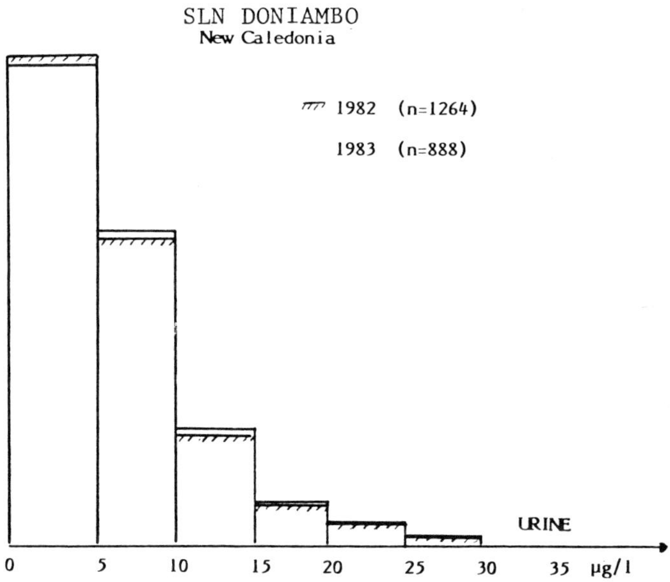

FIGURE 1. Relative urinary excretion of nickel by workers of the SLN plant in Doniambo, New-Caledonia, for the years 1982 and 1983.

In New Caledonia, there are no underground mines. In the open pit mines, urine nickel levels were the same as the controls in the general population. This is consistent with the ores being of low nickel content (2% to 3%).

In 1983, serum nickel analyses were performed by the IUPAC method (Brown et al., 1981) and by a direct method; the values observed were very low - not exceeding 6 µg/L.

However, for reasons of logistics, time and economic constraints, urine analysis appeared more reliable and more attractive for routine work in New Caledonia.

CONCLUDING REMARK

This preventive health program is in no way original but like others in the nickel industry SLN believes that biological monitoring is useful and should be carried out in addition to environmental monitoring.

~~To quote from Evert Nieboer's draft report on "The usefulness of Biological Monitoring"~~

1. The routine measurement of nickel in body fluids is technically feasible.
2. Nickel levels in body fluids correlate well with environmental exposure to soluble and insoluble nickel compounds.

3. Biological monitoring is a good adjunct to environmental monitoring in assessing the effectiveness of industrial hygiene measures designed to minimize exposure, but the establishment of threshold limit values with respect to these concentrations has little scientific justification.
4. The measurement of nickel body fluid levels may contribute significantly to knowledge regarding risk of adverse health effects attributable to exposure. These adverse effects could occur in fifteen to thirty years, and then, it will be important to know to what substances the patients were exposed.

REFERENCE

Brown, S.S., Nomoto, S., Stoeppler, M. and Sunderman, F.W.,Jr. (1981) Pure and Applied Chemistry, 53, 773-781.

461650

HEALTH SURVEILLANCE OF WORKERS EXPOSED TO NICKEL AND ITS COMPOUNDS

CAROLE MARY BISHOP
Health and Safety Executive,
25, Chapel Street,
London, NW1 5EE UK

INTRODUCTION

This paper firstly defines health surveillance, screening and biological monitoring, discusses fairly critically the methods available at present and then applies these to nickel exposed workers and suggests a basic scheme for Health Surveillance of workers exposed to nickel and its compounds other than nickel carbonyl. It is important that the divers - ity of exposures to nickel compounds should be given due consideration when devising health surveillance programmes.

Health surveillance procedures should not be offered merely as a "window dressing" that is as a substitute for primary technical control measures; it is an adjunct to control of risk. It is generally easier to institute a test or procedure than to withdraw it when its efficacy has been questionned, and the medical officer must consider carefully what benefit the individual worker or group will derive and what action on receipt of test results will be taken prior to instituting such procedures. An assessment of risk should be made within the plant and any Health Surveillance programme tailored accordingly. Health Surveillance should not be confused with epidemiological study and hypothesis testing.

GENERAL STRATEGY FOR HEALTH SURVEILLANCE

The purpose of routine health surveillance is to identify at as early a stage as possible any variation in the health of employees which may be related to working conditions. The individual or group should benefit by reduced morbidity or mortality from the procedures included in the programme. An assessment of the level of surveillance appropriate to the establishment should be made, taking into account all factors which may affect health and only those levels of surveillance which are relevent to the industry or processes should be adopted. A reassessment should be carried out as new surveillance methods become available. The surveillance programme may include initial and periodic examinations, supplemented by additional tests as appropriate. Good records are vital to all systems of Health Surveillance. These should be accurate and readily retrievable.

Screening is the presumptive identification of unrecognised disease or defect by the application of tests, examinations or other procedures which can be applied rapidly. Screening tests sort out apparently well persons who probably have early pathology or disease from those who do not, they are not intended to be diagnostic.

Biological monitoring is the measurement and assessment of workplace agents or their metabolites either in tissues, secreta, excreta, expired air or any combination of these to evaluate exposure and health risk compared to an appropriate reference.

For any Health Surveillance Programme it is necessary to know what changes may be expected to occur in exposed workers and with what frequency these may be expected. Can these changes be reliably detected and measured and what remedial action is available to reverse or arrest these changes? What resources are required to achieve this?

The author accepts the chronic effects due to exposure to nickel and its compounds detailed by Norseth (1983) namely:- an increased risk of respiratory cancer (lung, nasal cavity) among some groups of nickel workers, contact dermatitis and allergic asthma (reported from the plating industry following exposure to nickel sulphate). Local effects of nickel in the nasal mucosa have been subject to considerable clinical interest in a search for indicators of cancer development or precancerous lesions, these do not cause clinical symptoms. A number of other systemic effects of nickel have been demonstrated in experimental animals, but the results are of uncertain human significance because of dose level and route of exposure.

With regard to some of the screening procedures available, chest radiography, sputum cytology and nasal biopsy have been excluded from the proposed surveillance of nickel workers as they do not satisfy the criteria for a screening procedure (World Health Organisation 1982; Nickel Symposium, Lyon, 1983; Williams et al. 1982).

With regard to biological monitoring it was concluded at the the IARC Symposium on Nickel in the Human Environment Symposium Lyon (1983) that the determination of nickel in plasma and urine is feasible, but the significance of such measurements in relation to exposure and effects has not yet been established and that considerable efforts must still be made for the determination of speciation.

The decision as to whether to institute a biological monitoring procedure rests with the Medical Officer and if introduced it should be clearly indicated to the individual and workforce that at present laboratory estimations cannot be related to health effects and risk. Careful consideration should be given as to whether group results have any advantage over hygiene monitoring alone.

SUGGESTIONS BY THE AUTHOR FOR HEALTH SURVEILLANCE OF WORKERS EXPOSED TO NICKEL AND ITS COMPOUNDS

For the purpose of inclusion in any surveillance programme a "nickel worker" should be defined within the plant.

The initial assessment may include:-

- a detailed occupational history and a full clinical history with specific reference to skin conditions e.g. contact dermatitis and eczema and any upper or lower respiratory tract complaints
- a smoking history and a family history should be noted
- an assessment of the individual's ability to wear respiratory protection where this is necessary for certain operations should be carried out
- a baseline chest X-Ray may be considered useful
- skin testing may sensitise the individual to nickel so that detailed questioning is probably preferable
- where there is an increased risk of developing allergic asthma a validated respiratory questionnaire should be applied and baseline spirometry undertaken
- the inclusion of biological monitoring should be given very careful consideration by the Medical Officer prior to its inclusion as stated earlier in this paper.

The Periodic assessment may include:-

- the reporting of skin and respiratory complaints (and any condition the worker considers attributable to nickel exposure) should be encouraged
- an annual assessment of sickness absence records (with special note of skin and respiratory tract conditions) should be undertaken and any trends or patterns in sickness absence noted and investigated
- review of the individual on return to work after prolonged sickness absence and the individual assessed for fitness to return to work with nickel compounds
- individuals who develop contact dermatitis should be skin tested and if positive for nickel, consideration should be given to alternative employment where available

- where it is considered that an increased risk of allergic asthma exists, a validated respiratory questionnaire may be applied and periodic spirometry where indicated.
- an individual who requires further investigations should be referred to the family practitioner, with the individual's consent.

INFORMATION TO THE WORKER

The workers should be informed of the health risks and potential health risks associated with exposure to nickel and its compounds in that particular industry and trained in their role in preventive action. The individual should be informed of results of any tests undertaken and their significance explained to him.

RECORDS

Records identifying individuals should be retained by the employer. Individual job histories should be maintained using job groups which are as homogeneous as possible in terms of exposure to nickel and its compounds. Exposure data should also be recorded. Records should be maintained for at least 30 years. Where there is access to an Occupational Health Department details of absence due to sickness should be retained by the Occupational Health Staff and any referrals made including dates of medical examinations and any screening tests noted.

DISCLAIMER

The views expressed in this paper are those of the author and are not attributable to the Health and Safety Executive.

REFERENCES

Nickel in the Human Environment Symposium. Lyon 1983 Sponsors:- Ministry of the Environment, France, IARC, IPCS, IPCS, CEC, ILO, in press.

Norseth. T. 1983. Clinical Effects of Nickel in Nickel in the Human Environment (Symposium) Sponsors:- Ministry of the Environment, France, IARC, IPCS, CEC, ILO, in press.

Williams. D. E., Cortese. D. A., 1982. The present status of screening and diagnosis of early lung cancer in Seminars in Respiratory Medicine, 3 210-216

World Health Organisation (1982) Lung Cancer: epidemiology, treatment, prevention. Reappraisal of the present situation in prevention and control of lung cancer in Bulletin of the World Health Organization : 810-817.

STUDY OF NICKEL-BEARING DUSTS AND BIOLOGICAL IMPLANTS BY MICRODIFFERENTIAL THERMAL ANALYSIS

DANYS DEWALLY

Laboratoire de Chimie Minérale Appliquée
Université des Sciences et Techniques de Lille
59655 Villeneuve d'Ascq FRANCE

INTRODUCTION

The epidemiological studies carried out on workers in certain areas of nickel refineries have shown that a relationship exists between exposure to certain nickel compounds and cancer of the nasal passages and the lungs. F.W. Sunderman Jr. (1977) and R. Doll et al. (1977) have shown that in experimental animals various nickel compounds induce cancers at the site of exposure. Crystalline nickel sulfide particles (NiS, α Ni_3S_2) are especially potent carcinogens. This paper describes the microdifferential thermal analysis of pure α Ni_3S_2 (Heazlewoodite) under air pressure before and after tumorigenesis and the mechanism of oxidation of α Ni_3S_2. We try to apply these results to the dusts emitted by a matte roasting furnace. The microdifferential thermal analysis permits the analysis of nickel bearing dusts and biological implants.

MATERIALS AND METHODS

The Ni_3S_2 powder used is α Ni_3S_2 (Heazlewoodite). After tumorigenesis in rats or rabbits the Ni_3S_2 implants were generally recovered and studied. The reacted specimens were examined and characterized by X ray powder diffraction patterns on a Guinier de Wolff camera with a copper anode X ray (λK_{α} = 1,5406Å). The absence of nickel powder in Ni_3S_2 was determined by magnetochemistry with a Beauduin electromagnet and a Mettler balance. The microdifferential thermal analyser was a BDL instrument in which the sample and the reference material (Al_2O_3) were placed in a holder of 25 µl. They were then put in a specimen holder assembly in refractory material in a air or oxygen flow. The heating rate was 5 K/min, as measured by hollow platinel thermocouples which were welded on to the holders. The thermobalance was a Ugine Eyraud automatic recording balance. The heating rate was 2 K/min.

RESULTS AND DISCUSSION

Differential thermal analysis and thermogravimetric analysis curves of pure Ni_3S_2 powder (Heazlewoodite) in an air flow at atmosphere pressure are shown in Fig. 1.

Differential thermal analysis

The oxidation of nickel subsulfide in an air flow of atmospheric pressure shows three stages (DEWALLY and TRIDOT 1981). The first (T < 350 °C) corresponds to the formation of NiO and NiS according to the following mechanism-

$$Ni_3S_2 + O_2 \longrightarrow 2\ NiO + NiS_2$$
$$NiS_2 + Ni_3S_2 \longrightarrow 4\ NiS$$

$$Ni_3S_2 + 1/2\ O_2 \longrightarrow NiO + 2\ NiS$$

This reaction is marked by a wide exothermic peak.

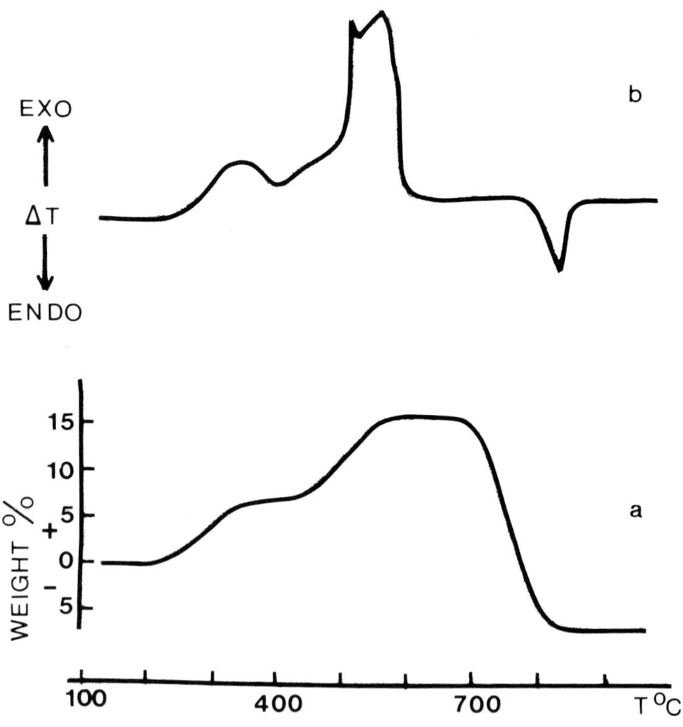

FIG.1. Ni_3S_2 in an air flow at atmospheric pressure: curves for (a) thermogravimetric analysis (b) differential thermal analysis.

The second (T between 350 °C to 600 °C) leads to the production of NiO and $NiSO_4$ and the emission of sulfur dioxide as follows-

$$3\ NiS + 9/2\ O_2 \longrightarrow 3\ NiO + 3\ SO_2$$
$$NiO + SO_2 + 1/2\ O_2 \longrightarrow NiSO_4$$
$$3\ NiS + SO_2 \longrightarrow 2\ NiO + NiSO_4 + 2\ SO_2$$

These reactions are marked by many strong exothermic peaks.

The third stage (T above 700 °C) results in the formation of NiO + (SO_2 + 1/2 O_2) from $NiSO_4$ decomposition. This reaction is marked by a endothermic peak.

Thermogravimetric analysis

On the TGA curve we observed three plateaux corresponding to the three reactions

$$Ni_3S_2 + 1/2\ O_2 \longrightarrow NiO + 2\ NiS$$
$$3\ NiS + SO_2 \longrightarrow 2\ NiO + NiSO_4 + 2\ SO_2$$
$$NiSO_4 \longrightarrow NiO + (SO_2 + 1/2\ O_2)$$

Differential thermal analysis of the implant is shown in Fig 2.

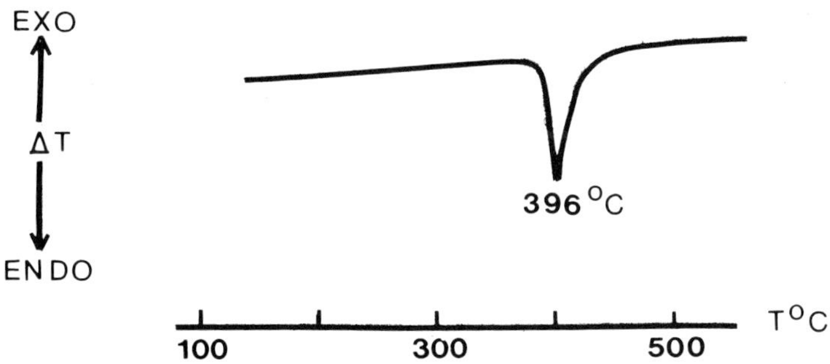

FIG. 2. Differential thermal analysis of the biological implant.

Dewally et al. (1980) have shown by X ray diffraction of a implant of Ni_3S_2 after tumorigenesis the presence of α Ni_3S_2 and rhombohedral β NiS (Millerite). By X ray diffraction it is impossible to detect very small quantity of β NiS produced concurrently with tumorigenesis: also we have developed a method of microdifferential thermal analysis (Dewally and Tridot, 1982). The DTA curve shows one endothermic peak at 669 K corresponding to the allotropic transition

$$NiS_{rhombohedral} \longrightarrow NiS_{hexagonal}$$

For the pure Ni_3S_2 there is no endothermic transition in this temperature range so DTA easily detects very small quantities of rhombohedral NiS produced by tumorigenesis. This fine detection is impossible by X ray diffraction. The transformation of Ni_3S_2 into NiS cannot be explained by the direct extrusion of nickel. This modification can occur however when sulfur is present. Thus it can be suggested that Ni_3S_2 is picking up sulfur from the host organism for the conversion into NiS.

CONCLUSIONS

These thermal microanalyses were realized in the laboratory in relatively the same conditions as these of the dusts emitted by a matte roasting furnace (Dewally 1983). Thus microanalysis allow us to identify the very fine particles of Ni_3S_2 that may enter the lungs of workers who may happen to be exposed to them. They also show that the Ni_3S_2 in its passage through the atmosphere of the workplace may transform itself into NiO or into other nickel bearing compounds. In summary, microdifferential thermal analysis permits the analysis of dusts and biological implants and detection of traces of compounds which are associated with biological reactions.

REFERENCES

Dewally, D. (1983). In : I.A.R.C. Monograph on Nickel in the Human Environment, International Agency For Research on Cancer, Lyon (in press).

Dewally, D. and Hildebrand, H.F. (1980). In : Brown, S.S. and Sunderman, F.W. Jr., eds., Nickel Toxicology, London, Academic Press, 51.

Dewally, D. and Tridot, G. (1981). In : Dollimore, D.(Eds) Proceedings of the Second European Symposium on Thermal Analysis, London, Heyden, 344.

Dewally, D. and Tridot, G. (1982). Abst. Eighth Annual Meeting I.U.P.A.C. Subcommittee on Environmental and Occupational Toxicology of Nickel, Dublin, 2.

Doll, R., Mathews, J.D. and Morgan, L.G. (1954). Brit. J. Ind. Med., 15, 102.

Sunderman, F.W. Jr. (1977). Ann. Clin. Lab. Sci., 7, 377.

GENERATION OF VARIOUS TYPES OF NICKEL CONTAINING AEROSOLS AND DETECTION OF NICKEL SPECIES IN THESE AEROSOLS

DIETER KLOCKOW and ROLF-DIETER KAISER
University of Dortmund, Department of Chemistry,
Otto-Hahn-Straße, D-4600 Dortmund 50, FRG

INTRODUCTION

For assessment of the ecological and toxicological risks of heavy metals in the environment, both the concentrations and the chemical forms of these elements are of special interest. In water analysis metal species may be identified by electrochemical methods (Florence and Batley, 1980). For investigation of sediments with respect to metal availability leaching procedures have been developed (Tessier et al., 1979). Metal speciation in air has so far been mainly restricted to volatile compounds by chromatographic separation techniques (De Jonghe and Adams, 1982). A few attempts have been made to achieve identification of some crystal structures in airborne particles using X-ray diffraction (Fukasawa, 1983).

The present study was designed to examine the potential of selective leaching procedures for speciation of nickel compounds in solid particles using artificially generated aerosols as test materials.

GENERATED AEROSOLS

The aerosols were generated by atomizing nickel salt solutions or a suspension of NiS in water. NiS was prepared by precipitation from homogeneous solution using thioacetamide.

Nickel formate aerosol was thermally decomposed to obtain Ni^{0} particles (Kanapilly et al., 1970). In order to generate NiO particles thermal decomposition of nitrate aerosol was employed.

CHARACTERISATION OF THE GENERATED PARTICLES

By using SEM (scanning electron microscopy) it was found that the generated particles were in the respirable range (optical count mode 1 - 2 μm for nickel sulfide, and 0.1 - 0.2 μm for the other compounds). Investigation by X-ray diffraction proved that the conversion products of the thermal decomposition of nickel formate and nickel nitrate aerosols were Ni^{o}- and NiO-particles, respectively. ESCA (electron spectroscopy for chemical analysis) studies have shown that the generated Ni^{o} was covered with an oxide layer similar to commercially available nickel powder. Identification of Ni^{o} below a NiO layer by ESCA is impossible because Ar^{+} etching not only removes the surface layer but at the same time reduces nickel oxide to metallic nickel (Kim and Winograd, 1974).

MIXING OF AEROSOLS

Using the experimental set up shown in Figure 1, it was possible to obtain binary mixtures of two different aerosols in a molar ratio between 1:10 to 10:1. In order to determine this ratio at least one of the aerosols should contain another

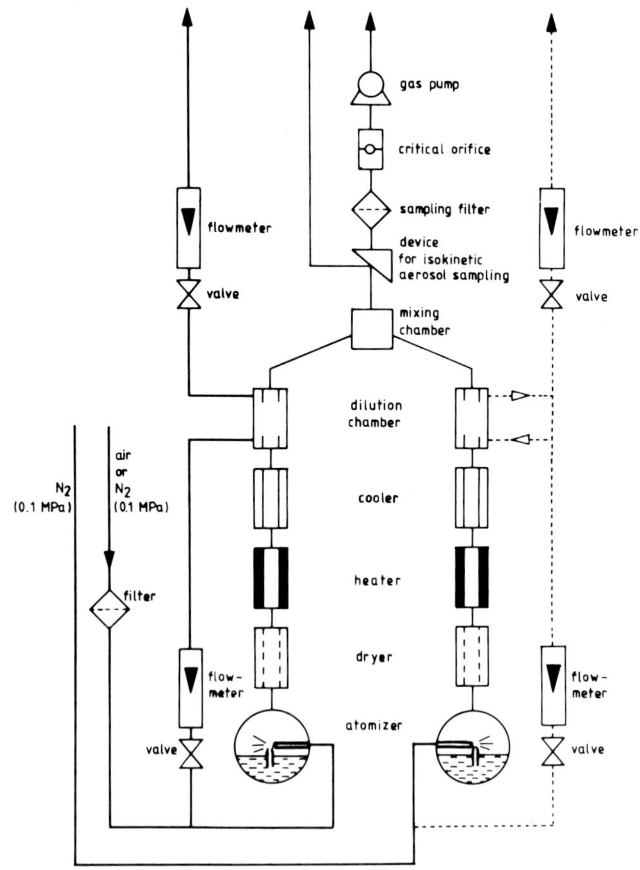

FIG. 1. Experimental set-up for the generation of binary mixtures of different aerosols.

element (Z > 15) as an internal standard which can be an anion or a cation added to one of the solutions to be atomized. After having collected the aerosol particles on a filter and having analysed this filter by X-ray fluorescence for Ni and the internal standard the ratio of the components of the binary mixture can be easily calculated.

SPECIATION BY RING OVEN TECHNIQUE

Extraction procedures have long been employed for phase analysis studies in metallurgical samples (Meineke, 1871). The separation of metallic and oxide phases in such samples is possible by selective dissolution of the metallic phase using $CuSO_4$ (Frerichs, 1908) or $KCuCl_3$ (Riott, 1941) solutions, respectively. In order to achieve a selective dissolution of nickel sulfide in the presence of metallic and oxidised nickel an aqueous solution containing hydrogen peroxide and di-ammonium hydrogen citrate was used (Katsnel'son and Osipova, 1960).

To apply these extraction procedures to separation of different phases in artificially generated aerosols and mixtures

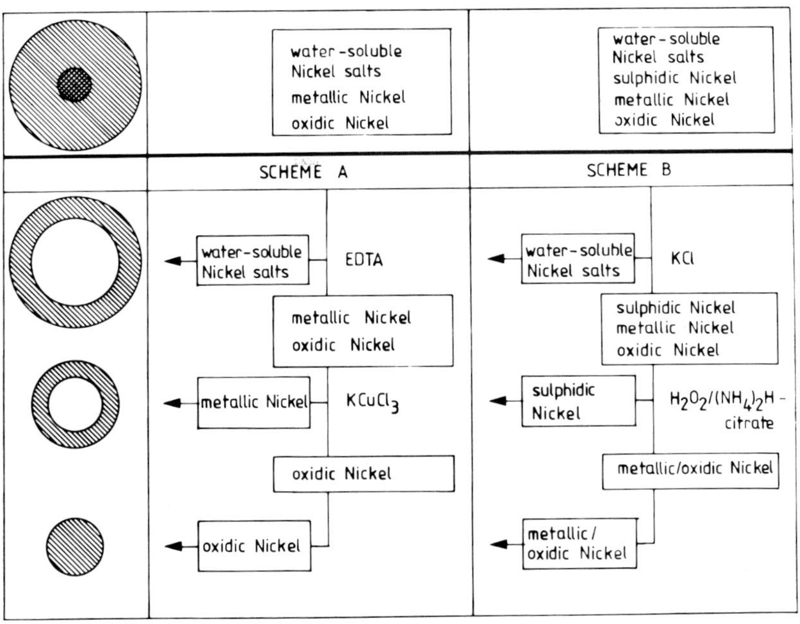

FIG. 2. Schemes for speciation of nickel-containing particles.

of aerosols the Weisz ring oven (Weisz, 1970) was used. This is a simple but highly versatile separation device, because it allows the different operations to be carried out directly on the collection filter with μg amounts of material. Depending on the reagent used for leaching the species either stay in the center of the filter or are concentrated in the "outer" or "inner" ring (Figure 2). The amount of Ni in these rings can be determined semiquantitatively by spot colorimetry or by flameless atomic absorption spectrometry, after dissolution in acid.

CONCLUSIONS

Using the Weisz ring oven as a simple microchemical separation technique it has been demonstrated that speciation of nickel in artificially generated particulate matter is possible by selective leaching. Preliminary studies have shown that the speciation scheme so developed should be applicable to the investigation of emissions from metal working industries and of dust in workplace atmospheres.

ACKNOWLEDGEMENTS

This work has been partially supported by the "Bundesanstalt für Arbeitsschutz", Dortmund, FRG.

REFERENCES

De Jonghe, W. R. A. and Adams, F. C. (1982) Talanta, 29, 1057 - 1067.

Florence, T. M. and Batley, G. E. (1980) CRC Critical Reviews in Analytical Chemistry, 9, 219 - 296.

Frerichs, G. (1908) Archiv der Pharmazie (Weinheim, Germany), 246, 190 - 205.

Fukasawa, T., Iwatsuki, M. and Tillekeratne, S. P. (1983) Environmental Science and Technology, 17, 596 - 602.

Kanapilly, G. M., Raabe, O. G. and Newton, G. J. (1970) Journal of Aerosol Science, 1, 313 - 323.

Katsnel'son, E. M. and Osipova, E. Y. (1960) Obogashchenie Rud, 5, 24 - 26; Chemical Abstracts (1961) 55, 26854.

Kim, K. S. and Winograd, N. (1974) Surfaces Science, 43, 625 - 643.

Meineke, K. (1871) Fresenius' Zeitung fuer Analytische Chemie, 10, 280 - 284.

Riott, J. P. (1941) Industrial and Engineering Chemistry, Analytical Edition, 13, 546 - 549.

Tessier, A., Campbell, P. G. C. and Bisson, M. (1979) Analytical Chemistry, 51, 844 - 851

Weisz, H. (1970) Microanalysis by Ring Oven Techniques, 2nd edition, Pergamon Press, Oxford.

NICKEL OXIDE AND NICKEL SUBSULFIDE RESULTING FROM ATTACK OF SUPER ALLOYS IN AERONAUTICAL GAS TURBINES BY FUEL COMBUSTION PRODUCTS

DANYS DEWALLY

Laboratoire de Chimie Minérale Appliquée
Université des Sciences et Techniques de Lille
59655 Villeneuve d'Ascq FRANCE

INTRODUCTION

Nickel is one of the main constituents of super alloys used for the production of aeronautical gas turbines. The vanes of these turbines are exposed to the attack of fuel combustion products which often contain sulfur and oxygen or sulfide compounds. High temperature corrosion of alloys in air or combustion gases which contain sulfur oxides may involve simultaneous oxidation and sulfidation. When this occurs reaction rates are considerably faster than in oxygen alone. The high temperature corrosion of metal and alloys in the atmospheres containing sulfur oxides may cause serious degradation of materials due to these simultaneous reactions. This has prompted great interest in studies of the reaction behavior of metals and alloys in gases such as SO_2, air + SO_2, etc. The reaction products have generally been reported to consist of NiO and Ni_3S_2.

DATA BASE AND PRINCIPAL FINDINGS

The nickel-oxygen-sulfur system

Rosenquist (1954) and Kullerud and Yund (1962) have shown that the nickel-sulfur system is relatively complicated (Fig. 1). The solid solubility of sulfur in nickel is relatively low and it is estimated to be less than 0,1 % by weight at 800°C. At higher concentrations Ni-S liquid solutions are formed and a eutectic exists at 635°C of composition between 20-23 % sulfur by weight. Nickel subsulfide (Ni_3S_2) is the most nickel rich sulfide. Liné and Laffitte (1962) and Liné and Huber (1962) have shown that this phase exhibits considerable non stoichiometry $Ni_{3\pm x}S_2$ at and above 550°C and it melts at 806°C. Ni_7S_6 is a stable phase below 573°C and above this temperature it decomposes to $Ni_{3\pm x}S_2$ and α $Ni_{1-x}S$.

The phases Ni_3S_4 (stable below 356°C) and NiS_2 are found at higher sulfur concentrations. Nickel monoxide NiO is the only stable oxide of nickel at high temperature. Nickel sulfate is stable up to 700°C.

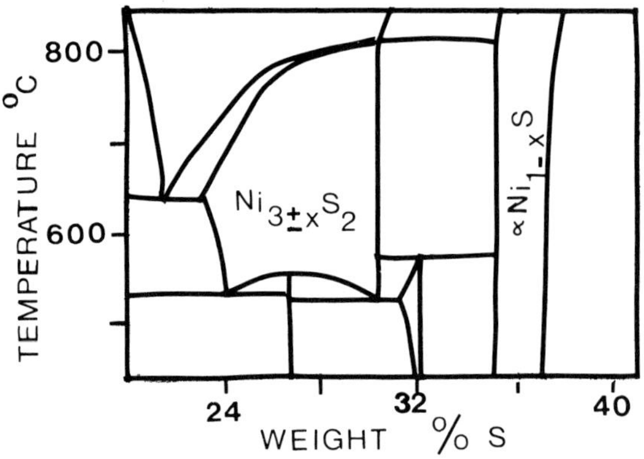

FIG. 1. Portion of the nickel-sulfur phase diagram.

Thermodynamic data and kinetic considerations

Subsequent discussion of the reaction behavior of nickel in SO_2 involves the thermodynamic properties of the Ni-O-S system. The following values of free enthalpies of formation and reaction for the various phases (expressed in J/mole) are used.

TABLE 1. Free enthalpy of reaction of the nickel-sulfur-oxygen system.

Reaction		ΔG°_T (J/mole)
$3Ni+S_2$	$\rightarrow Ni_3S_2$	- 331 223 + 163,06T
$Ni+1/2O_2$	$\rightarrow NiO$	- 244 321 + 98,44T
$3NiO+S_2$	$\rightarrow Ni_3S_2+3/2O_2$	401 739 - 132,26T
3NiS	$\rightarrow Ni_3S_2+1/2S_2$	107 426 - 52,63T
$NiO+1/2S_2$	$\rightarrow NiS+1/2O_2$	98 104 - 26,54T
$NiO+1/2S_2+3/2O_2$	$\rightarrow NiSO_4$	- 704 371 + 360,27T
$NiS+2O_2$	$\rightarrow NiSO_4$	- 802 476 + 386,82T
SO_2	$\rightarrow O_2+1/2S_2$	362 071 - 72,36T

The high temperature corrosion of nickel in SO_2 or $SO_2 + O_2$ has been studied by several investigators. The results reflect a complicated reaction behavior involving both oxidation and sulfidation. The reaction rates are much faster than in oxygen or air. The reaction kinetics may vary with temperature and exposure periods and are often interpreted in terms of linear and parabolic reaction kinetics or mixtures of these. A notable feature of the temperature dependence of the amount of reaction is that the reaction rate goes through maxima with increasing temperature. The temperatures at which the maxima occur differ somewhat in the different investigations. This maxima temperature was always between 600°C to 800°C. These rapid rates have often been explained by possible rapid diffusion of nickel through continuous nickel sulfide channels ($Ni_{3\pm x}S_2$) formed within nickel oxide scale. However the knowledge of the transport properties of β $Ni_{3\pm x}S_2$ phase is very limited. Rau (1976) and Lin et al. (1978) suggested that high temperature β $Ni_{3\pm x}S_2$ phase consisted of two different phases β_1 and β_2 $Ni_{3\pm x}S_2$. Yagi and Wagner (1982) have examined the composition relations of β $Ni_{3\pm x}S_2$ between 600 and 750 °C in order to test for the existence of the two phase regions of sulfur activity. The existence of β_1 and β_2 $Ni_{3\pm x}S_2$ was confirmed. The reaction of nickel with SO_2 in the temperature range 600-1100°C generally involved both oxidation and sulfidation.

This overall reaction behavior and the change in the mode of reaction with temperature and partial pressure of SO_2 may be conveniently analyzed by means of Ni-O-S stability diagrams in which the stability ranges of the different phases in the system are delineated. Such diagrams may be constructed in various ways by use of the thermodynamic data of the various phases in the Ni-O-S system. Isothermal diagrams are used in which the ordinate and abscissa denote the partial pressures of sulfur (S_2) and Oxygen (O_2) respectively. Such a diagram for a temperature of 600°C is shown in Fig. 2.

At temperatures lower than the melting temperature of eutectic Ni-Ni_3S_2 (635°C) two compact layers were detected : one internal layer of sulfide Ni_3S_2 and one external layer composed mainly of NiO with inclusions of Ni_3S_2 of geometry and dimension variable with temperature. At higher temperature these two layers become porous.

The rates of the reaction of nickel in these gas mixtures are extremely rapid.

For example Luthra and Worrell (1978) have shown that the corrosion rates of nickel in SO_2-O_2 atmospheres at 600°C are 10^5-10^6 times faster than the rate of oxidation of nickel in oxygen at 1 atm.

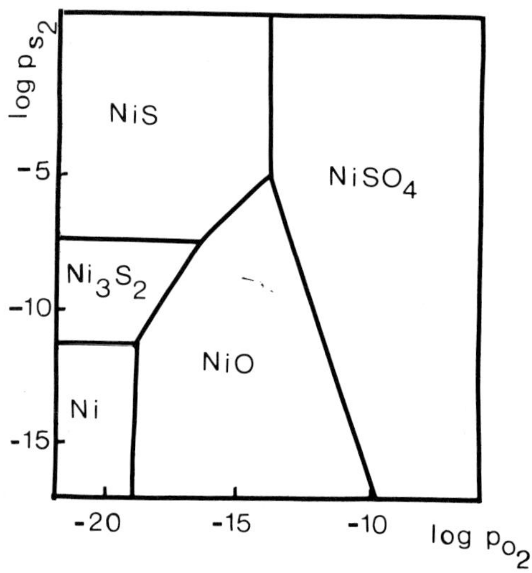

FIG.2. Isothermal diagram for the Ni-S-O system at 600°C.

CONCLUDING REMARK

Evidence for the existence of two high temperature phases $Ni_{3\pm x}S_2$ was examined. The findings show that to reduce corrosion and the formation of Ni_3S_2, it is necessary that aeronautical gas turbines should work with fuel containing as low a percentage of sulfur as possible.

REFERENCES

Kullerud, G. and Yund, R.A. (1962). J. Petrol., 3, 126.
Lin, R.Y., Hu, D.C. and Chang, Y.A. (1978). Met. Trans., 9 B, 531.
Liné, G. and Huber, M. (1963). C.R. Acad. Sc., C256, 3118.
Liné, G. and Laffitte, M. (1963). C.R. Acad. Sc., C256, 3306.
Luthra, K.L. and Worrell, W.L. (1978). Met. Trans., 9 A, 1055.
Rau, H. (1976). J. Phys. Chem. Solids, 37, 929.
Rosenquist, T. (1954). J. Iron Steel Inst., 176, 37.
Yagi, H. and Wagner, J.B., Jr. (1982). Oxid. Met., 18, 41.

SUMMARY REMARKS OF SESSION CHAIRMEN (Christian Bozec, Yngvar Thomassen, Jean-Paul Rigaut, J. Stuart Warner, Lindsay G. Morgan and Ernest Mastromatteo)

The papers in these sessions were concerned largely with the practical problems of assessing present or past occupational exposures to nickel-containing materials.

In view of the history of occupationally related disease at the Clydach nickel refinery in Wales, Dr. Morgan discussed the importance of monitoring the atmosphere for total dust and nickel compounds. Atmospheric monitoring, coupled with urinary nickel analysis, is a help in assessing the profile of exposure, but the role of urinary nickel in legislative control is debatable.

Dr. Roy focused on a NiPERA project which was designed to develop a consensus air sampling procedure which could be utilized by nickel producers to evaluate their workers' exposure to total airborne and respirable inorganic nickel particulates. The method recommended for use by NiPERA involves sampling for total airborne particulates with a filter of 0.8 μm mean pore size at a sampling rate of 2.0 L/min. The recommended procedure for respirable particulates involves sampling through a filter cyclone collecter at a similar rate.

Criteria and apparatus for a practical method of sampling airborne dust so as to determine occupational exposures were described by Dr. Wagg, and Dr. Courbon described a new device (MSA CIP 10) for sampling dusty atmospheres which combines sampling and pumping functions and appears to be compact and easily worn by workers.

A joint presentation of the results of determining atmospheric chromium and nickel in steel-making industries at various works in France and Germany was made by Dr. Klein and Dr. Gerling. Special attention was paid to the speciation of soluble nickel, total chromates and soluble chromates. The nickel levels appeared to be well within the present guidelines; greater interest attached to the findings for Cr(VI).

Dr. Thornhill described the approach made by NiPERA for the characterization of nickel compounds which are used in experiments *in vivo* or *in vitro*. The complicated case of "nickel oxides" was illustrated by examples of differing morphologies of this compound, depending on the manufacturing process.

Dr. Warner presented a detailed description of the material handling problems associated with the Copper Cliff sinter plant in Canada which operated between 1948 and 1963; here, nickel subsulphide was roasted to nickel oxide. Particular attention was drawn to the practice of recycling partially roasted material. The presentation included a method of estimating past exposures to airborne nickel compounds, which were predominantly nickel subsulphide and nickel oxide although some nickel sulphate may also have been present. The airborne dust also included carbonaceous materials derived from fuels and lubricants. These estimates clearly suggest that there were extremely high levels of nickel in the atmosphere (up to 300-400 mg/m^3) in the first few years of operation.

Dr. Warner's paper set the scene for Dr. Muir who showed how the very large epidemiological study carried out on Canadian nickel workers is suitable for detailed analysis of smaller groups. Moreover, investigation of the mortality patterns of sinter plant workers can show how the excess risk of the early period was considerably reduced when the dust levels were lower, although still very high by modern standards.

Prof. and Mrs. Goldberg reproted on-going epidemiological studies in New Caledonia. They and their teams were to be congratulated on having carried out such elaborate studies and reporting upon them so quickly. They had many problems in collecting adequate data in a developing country, and the researcher must judge for himself what weight can be put on the findings. The conclusion that "In New Caladonia it is probable that an increased risk did not exist, but this is not yet proven" is a very fair one. In this context, Dr. Morgan commented that at the time (10 years ago) of the preparation of the NIOSH Criteria for Occupational Exposure to Nickel there were no epidemiological investigations relating to mining or processing of lateritic ores. He himself had visited the old Le Havre plant of SLN in 1975 and considered that dust exposures there were sufficient to merit an epidemiological investigation; but since 'cause of death' remains confidential in mainland France, such a study was not practicable.

Thus the topics discussed by these speakers focused on the great difficulties in characterizing and estimating nickel concentrations in work place atmospheres. They addressed the need for establishing standard air sampling procedures. Most current sampling methods have been developed for sampling either total or respirable dust fumes. There is a great need for standardization, harmonization and quality control programs in the sampling and analysis of airborne nickel. Speciation studies are crucially important in assessing the risks associated with nickel exposures. The principles of proper health surveillance of workers exposed to nickel in its various forms are generally agreed: what is needed now are valid strategies and good tools for providing quantitative longitudinal supporting data.

INDEX OF CONTRIBUTORS

SUBJECT INDEX